Logic, Argumentation & Reasoning

Interdisciplinary Perspectives from the Humanities
and Social Sciences

Volume 7

Series editor
Shahid Rahman

Logic, Argumentation & Reasoning explores the links between Humanities and the Social Sciences, with theories including, decision and action theory as well as cognitive sciences, economy, sociology, law, logic, and philosophy of sciences. It's two main ambitions are to develop a theoretical framework that will encourage and enable interaction between disciplines as well as to federate the Humanities and Social Sciences around their main contributions to public life: using informed debate, lucid decision-making and action based on reflection.

The series welcomes research from the analytic and continental traditions, putting emphasis on four main focus areas:

- Argumentation models and studies
- Communication, language and techniques of argumentation
- Reception of arguments, persuasion and the impact of power
- Diachronic transformations of argumentative practices

The Series is developed in partnership with the Maison Européenne des Sciences de l'Homme et de la Société (MESHS) at Nord - Pas de Calais and the UMR-STL: 8163 (CNRS).

Proposals should include:

- A short synopsis of the work or the introduction chapter
- The proposed Table of Contents
- The CV of the lead author(s)
- If available: one sample chapter

We aim to make a first decision within 1 month of submission. In case of a positive first decision the work will be provisionally contracted: the final decision about publication will depend upon the result of the anonymous peer review of the complete manuscript. We aim to have the complete work peer-reviewed within 3 months of submission.

The series discourages the submission of manuscripts that contain reprints of previous published material and/or manuscripts that are below 150 pages / 85,000 words.

For inquiries and submission of proposals authors can contact the editor-in-chief Shahid Rahman via: shahid.rahman@univ-lille3.fr or managing editor, Laurent Keiff at laurent.keiff@gmail.com.

More information about this series at http://www.springer.com/series/11547

Matthias Armgardt • Patrice Canivez
Sandrine Chassagnard-Pinet
Editors

Past and Present Interactions in Legal Reasoning and Logic

Editors
Matthias Armgardt
Universität Konstanz
Konstanz, Germany

Sandrine Chassagnard-Pinet
Equipe René Demogue (CRD&P, EA 4487)
Université de Lille (UDSL)
Lille, France

Patrice Canivez
UMR 8163 Savoirs, Textes, Langage
Université Lille UFR Humanités
Département Philosophie
Villeneuve-d'Ascq, France

ISSN 2214-9120   ISSN 2214-9139   (electronic)
Logic, Argumentation & Reasoning
ISBN 978-3-319-16020-7   ISBN 978-3-319-16021-4   (eBook)
DOI 10.1007/978-3-319-16021-4

Library of Congress Control Number: 2015940408

Springer Cham Heidelberg New York Dordrecht London

Printed on acid-free paper

Springer International Publishing AG Switzerland is part of Springer Science+Business Media (www.springer.com)

Acknowledgements

All the chapters of this book originate from the French-German ANR-DFG project "JuriLog" (Jurisprudence and Logic) (ANR11 FRAL 003 01) between the University of Konstanz (Chair of Civil Law and History of Law, Prof. Dr. Matthias Armgardt) and the University of Lille (Chair of Logic and Epistemology, Prof. Dr. Shahid Rahman) with the support of the USR 3185 *Maison Européenne des Sciences de l' Homme et de la Société*, the UMR 8163 *Savoirs, Textes, Langage* and the EA 4487 *CRDP-Equipe René Demogue*. The project started in April 2012 and, since then, the members of the project have met twice a year at conferences which were held alternatively in Konstanz and in Lille. These opportunities were decisive for the taking shape of the works we are now presenting to the public. For this reason, besides the ANR and DFG, we would also like to thank the hosting institutions, namely the University of Konstanz, the University of Lille, the UMR 8163 *Savoirs, Textes, Langage* and the USR 3185 *Maison Européenne de Sciences de l'Homme et de la Societé*. We also express our gratitude to the organizers of the conferences, Bettine Jankowski (University of Konstanz), Sébastien Magnier and Juliele Sievers (University of Lille). Finally, our appreciation goes to Giuliano Bacigalupo for co-ordinating the different aspects linked to the preparation of this volume and to the various readers and referees for their silent help in selecting and improving the contributions to this volume.

Acknowledgement

Contents

Contributors

Matthias Armgardt is Full Professor of Civil Law, Ancient History of Law, Roman Law and Modern History of Private Law at the University of Konstanz, where he currently also holds the position of Vice-Rector of Academic Affairs. His research areas include Roman Law, Talmudic Law, Leibniz's legal philosophy, logic of law and European Private Law. He is a principal investigator of the Centre of Excellence "Cultural Foundations of Integration" at the University of Konstanz. Since 2012, he is co-directing together with Shahid Rahman (Université de Lille 3) the DFG-ANR research project "JuriLog" (Jurisprudence and Logic).

Full Professor of Civil Law, Ancient History of Law, Roman Law and Modern History of Private Law, Universität Konstanz, Konstanz, Germany

Giuliano Bacigalupo currently holds a Post-Doc position at the University of Konstanz in the ANR-DFG Project "JuriLog". He obtained his PhD in 2006 at the University of Bologna with a thesis on the philosophy of language of Edmund Husserl. Between 2009 and 2012, he was a lecturer at Seattle University (USA). His research focuses on the problems of philosophy of language and metaphysics, and, more specifically, on the notion of existence. Within the project "JuriLog", he works on the topic of legal fictions.

Fachbereich Rechtswissenschaft, Universität Konstanz, Konstanz, Germany

Rainhard Z. Bengez has been Institute of Advanced Studies' Carl von Linde Associate Professor of Mathematics and Models of Social Interactions at TU München until 2014. Since 2014, he is a visiting professor in Taiwan at the National University of Taipei and London School of Economics. He is an editor in legal theory and legal informatics for Jusletter IT and an associate editor of Oxford University Press' LPR – Law, Probability, and Risk Journal. Furthermore, he has been a special advisor to the EU Commission and the government.

MCTS, TU München, Carl von Linde Akademie Arcisstr. 21, 80333 Munich, Germany

Patrice Canivez is Full Professor of Moral and Political Philosophy, Director of the Department of Philosophy, Director of the Eric Weil Institute at the University of Lille 3 and member of the *Société Française de Philosophie*. Together with M. Crubellier, he is the editor of the series "Philosophy" at the Presses Universitaires du Septentrion (Lille). He has published books and articles on Aristotle, Kant, Hegel, Rousseau, Hannah Arendt, Paul Ricoeur, Eric Weil and on problems of contemporary political philosophy.

Full Professor of Moral and Political Philosophy, UMR 8163 Savoirs, Textes, Langage, Université Lille UFR Humanités, Département Philosophie, Villeneuve-d'Ascq, France

Sandrine Chassagnard-Pinet is Full Professor of Private Law at the Université Lille 2, co-director of the Centre René Demogue (CRD&P, EA 4487) and member of the team at the head of the Maison Européenne des Sciences de l'Homme et de la Société de Lille. Her research focuses on the theory of law, contract law and alternative dispute resolution. She has published books and articles on contractualization, globalization of the law and legal argumentation.

Full Professor of Private Law, Equipe René Demogue (CRD&P, EA 4487), Université de Lille (UDSL), Lille, France

Karlheinz Hülser has studied philosophy, theology, classics and sociology and has published mainly on L. Wittgenstein's earlier writings, on Plato and on Stoic dialectic. He taught ancient philosophy at the University of Jena and continues to teach it in Konstanz. Currently, he is studying the topic of friendship, retracing the influences of Stoic philosophy, particularly in the field of law, and exploring the influence of ancient logic in the sixteenth- and seventeenth-century Europe. Since 2012, he is engaged in the ANR-DFG Project "JuriLog".

Fachbereich Philosophie, Universität Konstanz, Konstanz, Germany

Bettine Jankowski is a PhD student in Law at the University of Konstanz, where she also holds a position as research assistant at the chair of Prof. Dr. Matthias Armgardt. Since April 2012, she is part of the research team of the ANR-DFG project "JuriLog" (Jurisprudence and Logic), where she is involved at the scientific as well as the organizational level. Her research focuses on the legal works of Gottfried Wilhelm Leibniz, especially his thesis on perplex cases.

Fachbereich Rechtswissenschaft, Universität Konstanz, Konstanz, Germany

Sébastien Magnier is a contractual researcher at the CNRS. He currently works on the ANR-DFG "JuriLog" Project and he is associated with Lille 3 University. He obtained his PhD in 2013. On the one hand, his research focuses on the possible links between epistemology and dynamic epistemic logics, and on the other hand, he uses dynamic logics in order to formalise some juridical notions, such as suspensive conditions. He has recently published papers on dynamic epistemic logics and their argumentative approach.

University of Lille 3, UMR 8163 Savoirs, Textes, Langage, 3 Rue du Barreau, 59650, Villeneuve-d'Ascq, France

Shahid Rahman is Full Professor of Logic and Epistemology at the University Lille 3 "classe exceptionelle (ech 2)" and member of the UMR 8163 *Savoirs, Textes, Langage*, where he is the director of the domain "Concepts and Philosophical Practices" and of the research topic "Logic and Argumentation". He is the editor of five book series for Kluwer-Springer and King's College and has published and edited several books on Logic, Epistemology and Philosophy of Science. In 2008, a book was dedicated to his work by King's College with, among others, contributions by J. van Benthem, J. Hintikka, G.S. Read and J. Woods. Prof. Rahman is also scientific director of the research project that federates 18 research laboratories and that is fostered by the MESHS-Nord Pas de Callais ADA (Argumenter, Décider, Agir). Since 2014, he is co-directing the ANR-DFG Project "JuriLog" together with Matthias Armgardt (University of Konstanz).

Département de philosophie, UFR Humanités, UMR 8163 Savoirs, Textes, Langage, Université de Lille 3, Villeneuve-d'Ascq, France

Juliette Sénéchal is Professor of Private Law at the University of Lille 2 and member of the European Law Institute. She has published books and articles on European contract law, on the impact of European law on French civil and commercial law, on law and environment and on law and management.

Equipe René Demogue (EA 4487), Université Lille Nord de France (UDSL), Lille, France

Juliele Maria Sievers is a researcher in the field of legal philosophy. She currently works at the University of Konstanz in the ANR-DFG Project "JuriLog". Her PhD thesis at the *Université Lille Nord de France* is about the logical aspects of Hans Kelsen's theory of norms. In 2011, she co-edited together with Prof. Dr. Shahid Rahman a collection of essays with the title *Normes et Fiction* (College Publications, Cahiers de Logique et Epistemologie, Vol. 11).

University of Lille 3, UMR 8163 Savoirs, Textes, Langage, 3 Rue du Barreau, 59650, Villeneuve-d'Ascq, France

Markus Winkler holds degrees in Mathematics from the Swiss Federal Institute of Technology and in Law from the University of Zurich. He has written on the influence of Greek logic on Roman jurists, in particular Salvius Julianus. His current research interests span Roman law and the history of German law in the seventeenth to nineteenth centuries.

Rechtswissenschaftliches Institut, Universität Zürich, Zürich, Switzerland

Chapter 1
General Introduction

Patrice Canivez and Giuliano Bacigalupo

Abstract This chapter provides a general introduction to the volume and highlights the interdisciplinary character of the research that led to this publication: all the contributions are the results of the bi-national Franco-German (ANR-DFG) project "JuriLog" (Jurisprudence and Logic), which foresees a strong collaborative work between jurists, philosophers and logicians. In addition, the chapter contains an overview of the topics addressed in the three parts of the volume: Roman law (Part I), Leibniz's writings on Law and Logic (Part II), and Current Interactions between Law and Logic (Part III). As the structure of the volume shows, the contributions have a strong historical and systematic focus. The chapter concludes by considering further developments of the project.

This volume presents the first results of an interdisciplinary research project on the relationships between logic and jurisprudence (JuriLog). Funded by the German Research Foundation (DFG) and the French Agence Nationale de la Recherche (ANR), the project is directed by Prof. Dr. Matthias Armgardt (University of Konstanz) and Prof. Dr. Shahid Rahman (University of Lille). The aim of the project is to explore the relations between legal reasoning and logic from a perspective both historic and systematic. The core idea of the project is to enhance the fruitful interactions between the two disciplines.

On the one hand, lawyers and legal scholars have an interest in emphasizing the logical character of legal reasoning. In this respect, the present enquiry examines the question of how logic, especially newer forms of dialogical logic, can be made fruitful as a significant area of philosophy for jurisprudence and legal practice. On the other hand, logicians find in legal reasoning a striving towards clear definitions and inference-procedures that is relevant to their discipline. In order to fully

P. Canivez
Full Professor of Moral and Political Philosophy, UMR 8163 Savoirs, Textes, Langage,
Université Lille UFR Humanités, Département Philosophie, Villeneuve-d'Ascq, France
e-mail: patrice.canivez@univ-lille3.fr

G. Bacigalupo (✉)
Fachbereich Rechtswissenschaft, Universität Konstanz, Konstanz, Germany
e-mail: giuliano.bacigalupo@uni-konstanz.de; giulianobacigalupo@googlemail.com

© Springer International Publishing Switzerland 2015
M. Armgardt et al. (eds.), *Past and Present Interactions in Legal Reasoning and Logic*, Logic, Argumentation & Reasoning 7, DOI 10.1007/978-3-319-16021-4_1

understand such reciprocal relationships, it is necessary to bridge the gap between law, logic and philosophy in contemporary academic research. The essays collected in this volume all work towards this common goal.

From a historical point of view, the awareness of the relationships between legal reasoning, logic and philosophy appears most clearly in Roman Law and Leibniz's writings. From a systematic point of view, the main point is that philosophers and legal scholars develop different logical-philosophical frameworks in order to address specific problems in law and the philosophy of right. At the crossroads of the two perspectives, the chapters of this book address such key topics as conditional legal acts, disjunctions in legal acts, presumptions and conjectures, conflict of values, Jørgensen's Dilemma, the Rhetor's Dilemma, the notion of legal fiction, and the categorization of contracts. The unifying problematic of these contributions concerns the conditional structures and more particularly the relationship between legal theory and legal reasoning in the context of conditions. Conditional clauses, conditional standards and legal fictions are at the heart of this interdisciplinary study of the relationship between law and logic.

According to this approach, the following chapters are arranged in three parts, dealing respectively with Roman Law, with Leibniz's writings on Law and Logic, and with current interactions between Law and Logic.

Part I. *Roman Law and Logic*. The second chapter "Proculus on the Meanings of *OR* and the Types of Disjunction" (Karlheinz Hülser) focuses on a crucial passage of the second book of Proculus's *Letters* where disjunctions and subdisjunctions are distinguished. The aim of the chapter is to interpret the passage on the background of Stoic approaches to the conjunction 'or'. The third chapter, "Disjunctive statements in Roman Legal Argument: Two Examples by Julian and Quintilian" (Markus Winkler), also dwells on disjunctions in a Roman Law context. The author illustrates how the Roman jurist Julian deals with an explicit disjunctive statement. Moreover, the chapter relies on a passage from Quintilian's *Lesser Declamations* to provide the missing elements of Julian's text and clarify the logical meaning of disjunction at stake.

Part II. *Leibniz, Law and Logic*. Matthias Armgardt's "Presumptions and Conjectures in Leibniz's Legal Theory" (Chap. 4) turns to the second historical focus of this volume and provides a first reconstruction of Leibniz's approach to presumptions and conjectures. The author considers three texts from different periods of Leibniz's production and establishes to what extent one can speak of a unified theory of presumptions and conjectures. Sébastien Magnier, in "Suspensive Conditions and Dynamic Epistemic Logic" (Chap. 5), offers a survey of Leibniz's approach to the notion of suspensive condition in law, thereby highlighting its epistemic and dynamic dimension. Bettine Jankowski's contribution, "The Rhetor's Dilemma: Leibniz's Approach to an Ancient Case" (Chap. 6), addresses a well-known juridical philosophical riddle. After an overview of the different attempts to solve the paradox, Jankowski shows how Leibniz provides the most convincing and conspicuous solution.

Part III. *Current Interactions between Law and Logic*. The seventh chapter, "The Epistemic Role of Dependent-Evidence and the Notion of Conditional Right", by

Shahid Rahman, studies the notion of conditional right by means of a dialogical approach to constructive type theory. The crucial insight developed by this contribution is that the dependency of the conditioned on the condition is defined with regard to the pieces of evidence that support the truth of the hypothetical, rather than the propositions that constitute it. In "Reasoning with Form and Content" (Chap. 9), Juliele Sievers and Sébastien Magnier put forward a logical frame that allows them to display the argumentative features behind legal decisions. To this end, the authors capitalize on the solution to Jørgensen's Dilemma provided by the legal scholar Hans Kelsen. In "Legal Fiction, Assumptions and Comparisons" (Chap. 8), Giuliano Bacigalupo develops a theory of legal fictions inspired by the philosopher Hans Vaihinger. Chapter 10, "Note on a Second Order Game in Legal Practice" (Rainhard Bengez), introduces a game theoretical framework, the P-game, to illustrate the process of practical decision making among lay judges and regular judges. This approach opens a fruitful theoretical framework for further investigations into negotiating strategies related to courts of lay assessors. Sandrine Chassagnard-Pinet ("Conflict of Norms and Conflict of Values in Law", Chap. 11) discusses the presence of antinomic rules in law, which seems to undermine the belief in the logical unity of the legal order. The author argues that a distinction between various categories of conflicts of norms paves the way to appropriate solutions for such antinomic rules. Finally, Juliette Sénéchal proposes a new categorization for service contracts in "Service Contracts: Problems and Methods of Classification" (Chap. 12).

Dedicated to the relationships between logic and jurisprudence, these contributions follow up on previous research published under the title *Approaches to Legal Rationality* in 2010 in the same series: Logic, Epistemology and the Unity of Science, directed by Shahid Rahman and John Symons (see [1]). In this former publication, researchers in logic, legal theory, ethics, political philosophy and computer science crossed the lines of their respective disciplines in order to open the path for new approaches to legal reasoning. *Legal Reasoning and Logic: Past and Present Interactions* develops the same interdisciplinary perspective while focusing on the aforementioned issues. In the coming years, the international research team set up to achieve this project will concentrate on the question of "the burden of proof". Further developments of the project will aim to clarify the rules of the burden of proof, of its distribution and possible reversal in the course of legal argumentation. A systematic grasp of these rules will contribute to a better understanding of legal practice and to the harmonization of judicial systems. Thus, the present volume is part of an ongoing and far-ranging research of both theoretical and practical significance.

Reference

1. D. Gabbay, P. Canivez, S. Rahman, A. Thiercelin (eds.), *Approaches to Legal Rationality* (Springer, Dordrecht, 2012)

Part I
Roman Law and Logic

Chapter 2
Proculus on the Meanings of *OR* and the Types of Disjunction

Karlheinz Hülser

Abstract From the second book of Proculus's *Letters*, the *Digestae* quote a passage that differentiates disjunctions and two kinds of subdisjunctions. The first part of this article deals with Proculus's person, his *Letters*, and with the juridical aims he pursued by referring to these logical distinctions. It provides the first premise for the later argument that the passage quoted in the *Digestae* is by far the oldest preserved document on Stoic subdisjunction(s), and that it is based on a Stoic textbook that is even older. The second premise is provided by a detailed documentation at the beginning of the second part of the article, to the effect that Proculus indeed relies on the tradition of Stoic logic. But the main theme of the second part of this article is its suggestion that we read the Proculus passage against the background of what else we know about the Stoic reflections on the conjunction "or". It is argued that the Stoic concept of disjunction was in a certain sense ambiguous and that Proculus and/or his immediate Stoic forerunners disambiguated it. Furthermore, a tendency is to be observed that might be seen as preparing the shift from the ancient concept of disjunction, which focuses on the exclusiveness of its components, to the more modern terminology that declares one of Proculus's "subdisjunctions" to form the now non-exclusive "disjunction". With this terminological development, Proculus and jurisprudence seem to have returned a stimulation to logic which they had received from just that discipline.

2.1 Introduction

During and after the second half of the second century BC, Roman jurisprudence was heavily affected by Greek scholarship, and emerged as a science through its influence step by step, in particular through the influence of Hellenistic dialectic. Since Fritz Schulz [25] has shown this so impressively in his *History of Roman Legal Science*, there has been little doubt that this was the case, though, of course,

K. Hülser (✉)
Fachbereich Philosophie, Universität Konstanz, Konstanz, Germany
e-mail: karlheinz.huelser@uni-konstanz.de

© Springer International Publishing Switzerland 2015
M. Armgardt et al. (eds.), *Past and Present Interactions in Legal Reasoning and Logic*, Logic, Argumentation & Reasoning 7, DOI 10.1007/978-3-319-16021-4_2

Schulz's presentation has met with some criticism and needs improvement. Yet his overall idea has been accepted, and many detailed studies have shown how the ideal of science took shape in Roman jurisprudence.

Proculus fits into this picture, too, an influential Roman lawyer who lived in the first century AD. Among the fragments of his *Letters*, however, there is one passage that is of interest not only with regard to the history of jurisprudence but also regarding the history of logic. It has been commented upon several times already, and more or less elaborately.[1] But it nevertheless remains worth the trouble of discussing the fragment again with regard to the meaning it has concerning the history of logic. That is what I want to do here.

The passage is conveyed to us in Emperor Justinian's *Digestae*. It is part of the chapter *De verborum significatione /On the meaning of words*, and reads as follows (D. 50. 16. 124 = FDS 978):

> Haec verba "ille aut ille" non solum disiunctiva, sed etiam subdisiunctivae orationis sunt. Disiunctivum est, veluti cum dicimus "aut dies aut nox est", quorum posito altero necesse est tolli alterum, item sublato altero poni alterum. Ita simili figuratione verbum potest esse subdisiunctivum. Subdisiunctivi autem genera sunt duo: unum, cum ex propositis finibus ita non potest uterque esse, ut possit neuter esse, veluti cum dicimus "aut sedet aut ambulat": nam ut nemo potest utrumque simul facere, ita aliquis potest neutrum, veluti is qui accumbit. Alterius generis est, cum ex propositis finibus ita non potest neuter esse, ut possit utrumque esse, veluti cum dicimus "omne animal aut facit aut patitur": nullum est enim quod nec faciat nec patiatur: at potest simul et facere et pati.
>
> The(se) words "So-and-so or So-and-so"[2] are not only disjunctive but also occur in subdisjunctive speech. We have a disjunctive [proposition] if, for instance, we say "either it is day or it is night", in which it is necessary that by accepting the one part the other one is rejected, and similarly by rejecting the other part the first one is accepted. Likewise an expression of similar formation can be a subdisjunctive [proposition]. Yet, there are two kinds of subdisjunctive [proposition]: one if because of the presupposed limitations it is as impossible for both to be the case as it is possible that none of them is the case; e. g. if we say "either he is sitting, or he is walking around". Since no-one is able to do both at the same time, it is possible however that a person does neither, e. g. a person who is lying down. The other kind is the case if because of the presupposed limitations it is as impossible for none to be the case as it is possible that each be the case, for instance when we say "every animal would be doing something actively or else it is suffering passively". For, there is no animal that could be neither active nor suffering; but it would be possible for an animal to be both at the same time, active and passive.[3]

Proculus certainly links up with the tradition of Stoic logic here when differentiating three ways to use the conjunction "or". What he discriminates looks very much like the contravalence, the exclusion, and the disjunction of modern

---

[1] See [1, pp. 221–223, 4, p. 110, 9, pp. 98–100, 11 = FDS 978/978a, 13, pp. 17–19, 15, pp. 82–88, 20, pp. 90–100].

[2] The Latin *Haec verba* is a plural form and rendered by "The(se) words". Yet the words in question form a formula (cf. below Sects. 2.2.3 and 2.3.3) so that in the present context the translations "The expression …" or "The formula …" might be considered perhaps better.

[3] The English translation is *mine*, only here and there adjusted to Samuel P. Scott's translation, Cincinatti 1932, as it is available in [17].

propositional calculus, though there are important differences, too. The terminology is somewhat confusing, since what nowadays is called a disjunction corresponds to the second type of Proculus's subdisjunctions; the first type corresponds to the modern exclusion, and his disjunction to our contravalence. Moreover, in explaining the three "or"s Proculus uses modalities – in contrast to propositional calculus. Nonetheless the similarity between Proculus's and the modern distinctions is impressive, and often enough it has been accentuated.

But the relevance of the passage with regard to the history of logic does not in fact parallel this similarity. In order to throw a different light on it, I shall first assemble some information on Proculus's life and his work. This gives us, on the one hand, a picture of how logical distinctions are used by Proculus for the purposes of jurisprudence, while on the other hand it establishes the first premise for a subsequent conclusion referring to the passage's relation to the tradition of Stoic logic. Further premises will be established by comparing Proculus's account more closely with what else we know about Stoic logic, in particular about the Stoic disjunction. I shall argue that the Stoic concept of the disjunction was in a certain sense ambiguous and that Proculus and/or the Stoic logicians he was relying on disambiguated it. To this extent, he/they not only presupposed but also modified the concept. This enabled them to develop clear-cut distinctions as they were submitted by Proculus.

2.2   On Proculus's Biography and on His *Letters*

In the *Digestae* the passage quoted above is preceded by a reference: originally it was written by "*Proculus libro secundo epistularum /* Proculus in the second book of (his) Letters". From this and from references related to some further paragraphs, it follows that the *Epistulae* formed a work consisting of at least eleven books,[4] and that they were still at hand to the editors of the *Digestae* in the sixth century AD. In addition the *Digestae* provide a little bit more information about Proculus, the author, by preserving (parts of) the history of Roman jurisprudence composed by Pomponius, a lawyer who lived in the second century AD.

---

[4]Unfortunately, Philipp Eduard Huschke in his collection of fragments on the *Iurisprudentiae anteiustinianae reliquiae* [12] omitted a chapter on Proculus. Only Otto Lenel in his *Palingenesia* arranged the material on Proculus into a kind of reconstruction: [18, vol. II, col. 159–184]. In view of a mistake in Cod. Florent. no. VI (see [18, vol. II col. 159 n. 2]) F. Schulz supposed that "the scribe must ... have mistaken *xii* for *iix*" and that the work consisted of 12 books: [25: 1961 p. 287[9]; 1946 p. 227[14]].

2.2.1   Proculus's Time and Identity

In this history, among other things, Pomponius describes the sequence of the
(generations of) lawyers since about the second century BC, and he accentuates
the excellent feedback Aulus Ofilius received around the end of the first century
BC. Two of his students were Ateius Capito and Antistius Labeo, who disagreed on
several points, so that for the first time one could speak of juristic trends or even
of juristic schools or sects. Capito's and Labeo's successors were – by this time
under Emperor Tiberius (14–37 AD) – Masurius Sabinus and M. Cocceius Nerva,
between whom the controversy concerning juristic orientation intensified. Masurius
Sabinus in turn was followed by Gaius Cassius Longinus, and M. Cocceius Nerva
by Proculus, whose names were used thereafter to signify the groups they presided
over; people called the one school the *Cassiani*, and the other one the *Proculiani*.
Cassius was succeeded by Caelius Sabinus and Proculus by Pegasus, both of
whom were active in the time of Emperor Vespasianus (69–79 AD) (D. 1. 2. 2. 53).
Pegasus was consul and *praefectus urbi* under Vespasianus, and also *praefectus urbi*
under Domitianus (81–96 AD).[5]

Besides describing Proculus's place in the sequence of notable lawyers, Pomponius characterizes the circumstances under which Proculus became head of the
group originating from Labeo. He writes (D. 1. 2. 2. 52):

> Nervae successit Proculus. Fuit eodem tempore et Nerva filius: fuit et alius Longinus ex
> equestri quidem ordine, qui postea ad praeturam usque pervenit. Sed Proculi auctoritas
> maior fuit, nam etiam plurimum potuit.
> Nerva was succeeded by Proculus. At the same time there was also Nerva, the son;
> there was still another Longinus belonging to the order of knights, who afterwards rose to
> the office of a praetor. Yet Proculus's authority was greater; for, he also had the greatest
> influence.[6]

Nerva died in the year 33 AD by committing suicide in Capri (see Tacitus, Ann.
6. 26. 1 f.; Cassius Dio Cocc. 58. 21. 4 f.). This, or his departure from Rome to
Capri, is the *terminus post quem* for Proculus's becoming the leading figure in
Labeo's tradition. But apparently Proculus was not the only candidate for Nerva's
succession. There were two rivals, Nerva's son and a further Longinus; to this extent,
Proculus's new position was not a matter of course, but possibly required some time.
However, according to Pomponius, Proculus's authority was greater – due to his
professional competence and due to other advantages, though it remains open just
where and how he achieved these. If indeed he was consul in the year 37 AD, as
R. A. Bauman assumes,[7] it will have been in the year 38 the latest when Proculus
became the head of Labeo's group.

---

[5]D. 1. 2. 2. 53; Juvenal, Sat. 4. 75-81. On Pegasus, see particularly [3, pp. 146–164].

[6]My translation, here and there adjusted to R. A. Bauman's passages on Proculus (see [3,
pp. 119–127]).

[7]See below.

In any case, Pomponius's account absolutely suffices for dating Proculus with certainty in the first century AD. Presumably born shortly before the Common Era, he took over the headship of Labeo's school towards the end of Tiberius's government, and died approximately towards the end of the rule of Emperor Nero (54–68 AD), though the time tolerances here are rather large.

This is, of course, only a small piece of information. For our purposes however it is an important one, and in this regard it suffices, though in other respects there are further important questions, at least two:

1. If we wish to know more about Proculus's person and biography, further investigations are required, since Pomponius gives no further identification by additional names or career details[8] so that the identity of Proculus is still unclear. The only remaining possibility is to check all bearers of the name "Proculus" who lived at the same time and come into question according to further biographical features. These investigations have been undertaken by various scholars, most recently by R. A. Bauman. He checked his predecessors's results, introduced new material into the debate, and discovered that Proculus was identical with Sempronius Proculus. He originated from Spain, from the same province as Seneca, the philosopher, though not from Cordoba but from Gades; in Rome he was, as it were, Nero's man. Furthermore, (a) this Sempronius Proculus from Spain was almost certainly the same person as Cn. Acerronius Proculus who was consul in the year 37 AD; and (b) certainly Sempronius Proculus was fully integrated into the Spanish population in Rome; to a certain extent he cooperated with Seneca, though without being Seneca's lawyer.[9]

2. Apart from Pomponius, not only Proculus's school but all juridical schools are mentioned also by other ancient writers, and the schools existed for more than 100 years. Most of the lawyers known to us can be assigned to one of them. Nevertheless, the information we have is somewhat confusing; and it is astonishing that, first, these schools are not named after their alleged founders but only after their third-generation heads, and, second, the account given by Pomponius features characteristics of an hellenistic historical construction. Therefore, and since there are additional confusing difficulties, the question is whether the schools were founded not by Capito and Labeo respectively but later by Sabinus and Proculus. In this case, we could say that the thirties of the first century were the years when Proculus became the head of his school – not so much due to inheriting the headship from Nerva but rather establishing and organizing the school of the Proculiani himself (on all this see [19], p. 197ff.).

---

[8]Besides Proculus this is the case only with Tuscianus: D. 1. 2. 2. 53; cf. [3, p. 119 f.].

[9][3, pp. 119–127, esp. p. 122 f. and p. 126 f.]. As to the results reached before Bauman, see [10]; see also [15, pp. 2–5].

2.2.2 Proculus's Letters

In order to date the passage from Proculus's *Letters* quoted at the outset, and to evaluate it with regard to the history of logic, in addition to a minimum of biographical data, it is crucial whether the *Epistulae* or at least this particular passage are authentic. Just this was doubted by F. Schulz, but convincingly defended by C. Krampe:

Considering a well-founded textual criticism, in the case of the *Epistulae* the methodological difficulty comes from the fact that there is nothing preserved from this work but what we find in the *Digestae*; there is no second independent textual tradition. Now, for a long time scholars acted on the assumption that the texts of the classical lawyers were modified, if at all, only by the editors of the *Digestae*. But against this, as a consequence of many detailed investigations, in the twentieth century a new working hypothesis developed, namely the supposition that in the sixth century the original writings of the classical lawyers were no longer available; what the editors of the sixth century had to hand would have been modified, post-classical new editions. Particularly in the case of Proculus's *Epistulae*, F. Schulz thought he could prove that what the editors had of the *Letters*, or at least of the passages passed down to them, was not the authentic version; and he offered our passage on the "or" as "a clear example" that "comes from neither Proculus nor the compilers, but from some academic person".[10]

C. Krampe developed an answer to this (see [15]). From the outset he was careful not to be influenced by dogmatic preconceptions, untested like the two traditional working-hypotheses mentioned above. The basic question he asked was about the literary form of the *Epistulae*. On the basis of certain passages he was able to determine this very form, and to show subsequently that all the other fragments fit this form extremely well, precisely on the supposition that the textual version passed down is still authentic and contains only very few editorial modifications that are identifiable with certainty.

In more detail, with respect to their literary form the *Epistulae* were structured into questions and answers, though in no case were they personal letters or occasional miscellaneous writings; they formed a systematic work. The *Epistulae* were teaching letters in which Proculus discussed juridical questions theoretically and developed well-grounded decisions for them. In the first century AD this literary form was no longer a new one but may be considered a topical adaption of the older dialogue form that had in antiquity become famous through Plato's work in particular; in Rome it was brought back to attention especially by philosophers. Regarding the subsequent structure, Proculus predominantly used dihaeretic procedures that were similarly old, and likewise became common in Rome through hellenistic dialectic. Such procedures consist in differentiating genera, according

---

[10][25: 1961, p. 287; 1946, p. 227]. Whom Schulz [25: 1961 p. 287[11]] names "some academic person" here, in his first version in German he calls a "Scholastiker". See also the summary by Krampe [15, p. 10].

to appropriate properties, into species, so that a conceptual structure emerges that may be used to systematize as well as to solve a problem. According to Krampe's analyses, a whole slew of Proculus fragments has to do with such differentiations. These fragments still indicate that Proculus appears to have developed his dihaeretic distinctions in various ways and to have used them for the analysis or modification of an issue on the one hand, and for concept formation on the other hand; in addition to systematisation, therefore, they were also used for the solution of the problems under discussion.

According to these considerations, the *Epistulae* formed a very scholarly work, a work that in a distinguished way instantiated the transformation of the Roman jurisprudence into a science modeled on Greek examples. There we find the disposition of Stoic dialectic implemented, even though this level of reception of Stoic logic is not the one we are seeking here. We are interested particularly in Proculus's explanations regarding the "or".

As pointed out above, *all* fragments unconstrainedly fit into the overall picture developed by Krampe; and if additionally the passages are subjected to individual examination with regard to textual criticism, as Krampe did, then in no case there is any serious reason to doubt their authenticity. This result explicitly applies also to our fragment from the second book of the *Epistulae*, also examined by Krampe [15, pp. 82–88]. Moreover, Krampe was able to clarify the relation between this fragment and a certain constitution in the *Codex Iustinianus*, and thus pointed out the juridical interest Proculus took in these logical differentiations.

2.2.3 The Juridical Interest in the Three Types of ›or‹

In an unremarkable footnote, Otto Lenel has already pointed out a certain *constitutio* by Emperor Justinian, and this reference has been repeated by other scholars on various occasions.[11] The reference is to C. 6. 38. 4, a constitution that is also about the expression "ille aut ille" ("So-and-so or So-and-so"). It reads as follows[12]:

> *pr.* Cum quidam sic vel institutionem vel legatum vel fideicommissum vel libertatem vel tutelam scripsisset: "ille vel ille heres mihi esto" vel "illi aut illi do lego" vel "dari volo", vel "illum aut illum liberum" vel "tutorem esse volo" vel "iubeo", dubitabatur, utrumne inutilis sit huiusmodi institutio et legatum et fideicommissum et libertas et tutoris datio, an occupantis melior condicio sit, an ambo in huiusmodi lucra vel munia vocentur et an secundum aliquem ordinem admittantur, an uterque omnimodo, cum alii primum in institutionibus quasi institutum admitti, secundum quasi substitutum, alii in fideicommissis posteriorem solum accepturum fideicommissum existimaverunt, quasi recentiore voluntate testatoris utentem.

---

[11]In addition to Lenel [18, vol. II, col. 161], see esp. [21, p. 99 f.].

[12]Latin text according to [16]. English translation by S. P. Scott (cf. above note 3); textual completions *mine* (in square brackets).

1. Et si quis eorum altercationes singillatim exponere maluerit, nihil prohibet non leve libri volumen extendere, ut sic explicari possit tanta auctorum varietas, cum non solum iuris auctores, sed etiam ipsae principales constitutiones, quas ipsi auctores rettulerunt, inter se variasse videntur. Melius itaque nobis visum est omni huiusmodi verbositate explosa coniunctionem "aut" pro "et" accipi, ut videatur copulativo modo esse prolata et magis sit παραδιάζευξις, ut et primam personam inducat et secundam non repellat. ...
2. Sin autem una quidem est persona, res autem ita derelictae: "illam aut illam rem illi do lego", vel "per fideicommissum relinquo", tunc secundum veteres regulas et antiquas definitiones vetustatis iura maneant incorrupta, nulla innovatione eis ex hac constitutione introducenda.
3. Quod etiam in contractibus locum habere censemus.

pr. When anyone appoints an heir, leaves a bequest, creates a trust, makes a grant of freedom, or establishes a guardianship, in the following words: "Let either So-and-so, or So-and-so be my heir," or "I give and bequeath to So-and-so," or "I wish property to be given to So-and-so," or "I desire that So-and-so, or So-and-so, shall become free, and act as guardian," or "I order this to be done," a doubt arose whether the appointment, the bequest, the trust, the grant of freedom, or the appointment of a guardian made in this way was not void; and whether the position of the party in possession was the better; or whether both parties were called to enjoy or assume benefits or burdens of this kind, and whether they should be admitted to any order, or whether both should be admitted without distinction. In the case of the appointment of heirs, some authorities thought that the first person named should be considered as the designated heir, and the second as the substitute; and others held that in the case of trusts, only the last person mentioned would have the right to accept it, as availing himself of the final intention of the testator.

1. Anyone who desires to succinctly dispose of the disputes of these jurisconsults will have no insignificant number of volumes to examine, as there is a great variety of opinions to be reconciled, for not only the legal authorities, but also the Imperial Constitutions which the said authorities have cited, are known to differ. Therefore having rejected all this verbosity, it has seemed to Us preferable that the conjunction "or" should be taken to mean "and," so that it may be understood in a certain sense to be copulative [and rather is a *paradiazeuxis* / subdisjunction], and hence [may] admit the first person mentioned without excluding the second; ...
2. Where, however, only one person is mentioned, but property is left as follows, "I do give and bequeath such-or-such property to So-and-so," or "I leave it to So-and-so in trust," then, in accordance with the ancient regulations, and the provisions of antiquity, the laws remain unimpaired, no change having been introduced in them by this Constitution.
3. We order that this rule shall also apply to contracts.

At the outset the constitution suggests that the expression "ille aut ille" is frequently used in testamentary dispositions, intended to alternatively appoint several persons as heirs, leave a bequest, create a trust, make a grant of freedom, or establish a guardianship; the classical lawyers had doubts how to interpret these testaments, and they devised several suggestions, three or five, depending on the mode of reckoning. The case was disputable and the discussion controversial. And now, i. e. in the sixth century, Justinian solves the problem by establishing a fixed rule of interpretation. The rule conforms to one of the suggestions, and says that in these and only these cases the testament is to be referred to both of the persons mentioned, or to all of them respectively. To explain the rule he adds that the word "aut" must be understood in the sense of "et" ("and") here, hence in a certain sense as a conjunction, better: as a subdisjunction.

Hence Justinian's constitution obviously refers to the Proculus passage, and *ex post* it clarifies the juridical purpose of these *prima facie* purely logical remarks of Proculus about "or". Proculus used them to establish his suggestion how to interpret the expression "ille aut ille" in testamentary dispositions, at least inasmuch as questions of heritage, bequest etc. are concerned.

As a matter of fact, however, it would be impossible to recognize this juridical purpose solely from the *Epistulae* fragment; it can be determined only by means of the *Codex Iustinianus*. In our fragment, the original context is omitted, and even the structure of question and answer is lost. What remains is nothing more than a dihaeretically structured account of three types of "or". For these oddities, Krampe offered a convincing explanation that at the same time shows how the authenticity of D. 50. 16. 124 is confirmed by C. 6. 38. 4: "Through its standardised authoritative solution … this constitution rendered any discussion concerning the problem of the *alternatio personarum* in relation to special cases superfluous. Therefore the editors borrowed from Proculus's work only the minimum into the *Digestae* that is required to justify the constitution linguistically, and significantly they did so under the same title *"De verborum significatione"* under which the constitution is subsumed in the codex" [15, p. 87].

2.3   The Aspect of the History of Logic

2.3.1   The Proculus Fragment as the Earliest Testimony Concerning Stoic Subdisjunction

The Proculus fragment on the three kinds of "or" testifies to a utilisation of logic for jurisprudence. After having observed the juridical aspect of this incident we can now go on to carve out the aspects concerning the history of logic.

First of all, we have to record what is most obvious and has been accentuated again and again: the passage from Proculus forms a consistent account of those three propositional connections by means of "or" that are known in modern propositional logic, too, though his "or"-connections do have different names and deviate from the modern ones in other respects.[13] More precisely, Proculus ties in here with the tradition of *Stoic* logic. The evidence for this specification is overwhelming.

Proculus's Stoic background can be inferred from the fact that no other ancient logical tradition is known where similar considerations on propositional conjunctions were cultivated. To the Stoics, then, there is no alternative; their logic is the only candidate here. A further unmistakable feature is that – except, of course, in the case of the juridical standard expression "ille aut ille" – Proculus meticulously obeys the Stoic rule that in a compound proposition the conjunction must be specified

---

[13] As will be seen below, the main difference is that the Stoic propositional connections allow for more than merely two components of a (sub)disjunction.

before *every* partial proposition, i. e. not only between the components but also before the very first one. In the next section, I will come back to this topic. Third, Proculus is using Stoic standard examples. In particular the day/night example is to be found in many fragments on Stoic logic, not least in the very important accounts by Diogenes Laertius and Sextus Empiricus.[14] And it is used here also in the same way, i. e. "It is day" and "It is night" are considered to be mutually exclusive; when connected by "either ... or ... " they form a disjunction, and we have to understand that precisely one of the constituents is true.

Further characteristics of a Stoic logical school-tradition in Proculus's sentences are: the Stoics accounted in the first place for what they called the "disjunction" (διεζευγμένον ἀξίωμα): just that "or" that announces that precisely one of the parts connected by it is the case while all the other parts are not the case. The testimonies we have even suggest that this exclusive "or" was the only "or" determined more precisely by Chrysippus.[15] At least it seems that the explanation of this "or" was considered rather simple. However, this same "or" forms the starting point for Proculus, too. Only after having explained the exclusive and exhaustive "or", the Stoics differentiated from it what they called "subdisjunction(s)", and so did Proculus. How he proceeded is shown by the passage quoted above; and how the Stoics proceeded is sufficiently clear from remarks by Gellius, Galen and Apollonius Dyscolus respectively.[16] Finally, there is a terminological argument. According to Gellius the Latin word "disiunctum" translates the Greek expression διεζευγμένον ἀξίωμα; analogously a παραδιεζευγμένον ἀξίωμα is rendered in Latin by "subdisiunctum". Proculus and the other sources mentioned use just these Latin terms. The corresponding Greek terms were typical Stoic expressions that in Galen's, Gellius's and Apollonius Dyscolus's time (second century AD) were accepted widely. But originally this terminology competed against the Peripatetic terminology,[17] and perhaps it still did in the first century AD. Proculus, however, apparently continued the terminological customs of the Stoics.

At the same time these details show that Proculus is building on a logical textbook by the Stoics. This is suggested on the one hand by his examples. The day/night example has already been mentioned. And the example of the various activities of a human being (walking, sitting, standing, lying, etc.) also occurs in Gellius (Gellius NA 16. 8. 14 = FDS 976). On the other hand it follows from the clarity and circumspection, the distinct symmetry and stylistic aplomb by which

---

[14]See in particular Diogenes Laertius 7. 72, 81 (= FDS 914, 1036); Sextus Empiricus, AM 8. 95; PH 2. 158, 201 (= FDS 915; 1128, 1135).

[15]Even Chrysippus's *Logical investigations* do not point to a different type of "or": cf. FDS 698, or see the more recent edition of the papyrus by Livia Marrone [20].

[16]For these testimonies, see FDS 976 f. and 979. For the Gellius passage see also the respective quotations in the Sects. 2.3.3 and 2.3.4 of this paper.

[17]As is to be seen from Galen's *Institutio logica* and from Alexander's of Aphrodisias *Commentaries* on Aristotle, instead of διεζευγμένον the Peripatetics originally preferred the term διαιρετικόν. Much evidence for this was collected by M. Frede [9, p. 93 n. 20]. In FDS, see e. g. 977.

Proculus explains the two subdisjunctions. All this requires a logical textbook as a model, according to which Proculus could arrange his text. He will not have simply copied it, since what Gellius says about Stoic disjunctions and subdisjunctions – to be quoted in part below – seems to be nearer to an original Stoic account than Proculus's report.[18] But a presentation like that by Proculus is generated best on the basis of a logical textbook of high quality. Finally, Proculus has no use for the Stoic precursor of Sheffer's stroke; he needs it only as a quasi-rhetorical means in order to describe the other subdisjunction convincingly, i. e. the non-exclusive "or" he is interested in. To do this presupposes a scholarly fundament, in other words: a logical textbook, perhaps one that Seneca pointed out to Proculus.

Armed with these results and with what we know about Proculus's life and his *Epistulae*, we are in a position now to draw three noteworthy consequences, two with regard to the history of logic and one regarding interpretation:

1. Proculus refers to the tradition of Stoic logic, lived during the first century AD, and embarked on the most productive period of his life probably in the thirties of that century. Given this, and given the fact that all other testimonies we have regarding Stoic subdisjunction(s) stem from authors who lived in the second century AD or even later,[19] it follows that the Proculus passage is by far the earliest document referring to Stoic subdisjunction(s), more than 100 years older than all the other testimonies.[20] Furthermore, it shows quite reliably that the Stoic logicians discussed the topic during the early first century AD (and that they did so even if the subdisjunction(s) were not yet a standard element of the school's logic).

2. Presumably, however, the Stoic logicians had started these investigations very much earlier. Otherwise, they would not have had enough time to develop their distinctions to such an extent and to record them in textbooks in a way that Proculus could develop. At what time they really began these discussions is difficult to estimate. In Cicero's *Topica*, dating from 44 BC, these or similar differentiations of the "or" are still missing.[21] But this does not mean much. In reality the beginnings of this debate will have gone back significantly into the Hellenistic period, though we do not know what precisely was discussed at that time, or how, or by whom.

3. In spite of the unambiguous marks of Stoic logic in Proculus's account, we must also register minor and major differences. The most important one is that, in order

---

[18]Notice that Proculus does not mention "conflicts" between the constituents of disjunctions and subdisjunctions which concept is crucial in the Gellius passage. Moreover, while Proculus describes the various "or"s by juxtaposing the respective criteria, Gellius approaches subdisjunctions explicitly by weakening the criteria for disjunctions.

[19]Cf. FDS 973 ff. Diogenes Laertius's report on Stoic logic does not mention subdisjunction(s); cf. FDS 914.

[20]This has been observed already by Frede [9, p. 98]; see also [4, p. 110 n. 88].

[21]When reporting a Stoic list of seven (!) indemonstrable hypothetical arguments, Cicero does not say anything about any kind of subdisjunction; see *Topics* 53–57 = FDS 1138.

to describe the disjunction and the subdisjunctions, Proculus uses modalities while other sources, for instance Gellius and Galen, refer to incompatibilities. Because of the close relationship between Proculus and the tradition of Stoic logic those differences will presumably not be too important. Regarding their interpretation, therefore, it will be advisable to let them appear as small as possible, i.e. not greater than essentially necessary.

With these consequences in mind, we return to Proculus's text and pick up a different thread.

2.3.2   The Palette of ›Or‹-Conjunctions in General

Proculus wanted to answer a juridical question; in view of certain juridical circumstances he needed a non-exclusive "or". For those purposes, then, he exploited Stoic logic and was quite successful as far as jurisprudence is concerned; in the sixth century his proposal was officially and permanently accepted. But what happened to Stoic logic in this exploitation? Sometimes Proculus's account is compared to modern propositional logic and by its conciseness shows a systematic closeness.[22] All the same, we may ask whether this closeness presupposes a kind of filtering the preceding Stoic logic, and whether Proculus or the Stoic textbook he used put something aside to develop that account. Or were anterior Stoic reflections on "or" even modified by him or his immediate Stoic forerunners?

Simplicius notes quite generally that regarding the conjunctions ἤ and ἤτοι (both: "or") the dialecticians accounted for many distinctions (Simplicius, *In Arist. Categ.* p. 25, 13 f. = FDS 972). In fact, beyond the picture outlined by Proculus, the Stoics also reflected upon an "or" they termed declaratory (διασαφητικός). Concerning this "or", there are only a few fragments, and one of them additionally says that the Stoics even identified, as a special type of the declaratory "or", an elenctic "or". This fragment is the passage from the *Epimerismi Homerici* already mentioned. According to the editor, A. R. Dyck, these epimerisms follow the tradition of Apollonius Dyscolus and Herodianus, and were preserved for us probably by Choeroboscus.[23] Accordingly, the fragment originated rather late. As an example for the elenctic "or" we are offered a verse from the Iliade: "βούλομ' ἐγὼ λαὸν σόον ἔμμεναι ἢ ἀπολέσθαι / for I would have the people live, ē (or, not) die" (Il. A 117; transl. S. Butler). Apparently, the "or" is elenctic here, i.e.

---

[22]For instance, Krampe [15, p. 85], stressed this and used it as an argument to defend the text's authenticity.

[23]See A. R. Dyck (ed.), *Epimerismi Homerici*, vol. 2, Berlin/New York 1995, p. 355 (η 20), and Dyck's introduction, esp. p. 23 f.; if Choeroboscus lived after Joannes Damscenus (*ca. 650), he presumably lived in the eigth century. FDS 980 exhibits the passage on the conjunction η according to C. A. Cramer's edition, Oxford 1835.

refuting, since what follows actually is refused; for, precisely what the speaker, in this case Agamemnon, says in the preceding part of the proposition, is what he unambiguously wants: that "the people live".

Unlike the elenctic "or" the standard case of the declaratory "or" is rather comparative and expresses a preference: "More it is day ἤ/ē/than it is night"; "Less it is day ἤ/ē/than it is night". While in English the two propositions are connected by "than" here, the Greeks used the very same word as they did to express a disjunction or a subdisjunction: ἤ/ē. For this reason the Stoics discussed the declaratory "or" in the context of propositional logic, particularly in connection with the disjunction. The Greek "ἤ/ē" marks the beginning of the second part of a non-simple proposition, and in a standard-declaratory proposition the associated first part begins with "more" or "less". For symmetry reasons the Stoics considered these words among the conjunctions.[24]

In a Latin linguistic environment there were no parallels to these types of "or". Moreover, neither these "or"s nor the respective non-simple propositions were of any interest for Proculus's purposes. Thus, in his account there are no traces of these Stoic reflections. Though no-one would blame him for any omission here, the observation confirms that Proculus exploited Stoic logic for his own juridical purposes, and extracted what he was interested in from its original context. His material thereby lost its vicinity to the declaratory and to the elenctic "or", and one wonders whether it underwent other slight modifications as well, those referring to the logically relevant core area of the "or"-propositions. To provide an answer it will be appropriate to give a more detailed description of what a disjunction is according to the Stoics; from there we shall return to the subdisjunction(s).

2.3.3 On the Disjunction in Stoic Dialectic

Proculus is interpreting the expression "ille aut ille", which is a formula and has a certain function in everyday life.[25] It contains the particle "aut" ("or") and thereby provides the opportunity to use Stoic logic for the purpose of interpretation, despite of a certain difference:

In the formula the particle appears only *once*, namely between the disjuncted components, and not additionally also at the beginning of the disjunction. The Stoic rules requested this intrinsically, and Proculus knew that they did: in the text that follows the formula, he regularly writes "aut ... aut ...", in his final example also "nec ... nec .." and "et ... et ...". In Greek, since the time of Chrysippus the Stoics even considered it important that the "or" in front of the very first part of

---

[24] Regarding the declaratory proposition cf. Apollonius Dyscolus, *De coni.* p. 222 f. = FDS 981, the passage from the *Epimerismi Homerici* mentioned above, and last but not least Diogenes Laertius 7, 72 f. = FDS 914.

[25] See above Sect. 2.2.3; cf. also D. 32. 25.

the disjunction should not be a simple ἤ but an ἤτοι, so that the entire disjunction had the form ἤτοι ... ἤ ... ("either ... or ..."). Regarding the negation, the conjunction, and the conditional, the Stoics had similar rules in their dialectic; the conventions concerning the declaratory proposition were already mentioned. By this standard, ambiguities were avoided or else removed, and it was ensured that the very first word of a proposition already determines the proposition's logical form, or at least that the scope of the various conjunctions was clear, similarly to what modern logicians achieve by brackets or by using the Polish notation.[26]

More important than the syntactical rule is the fact that a disjunction can and often will or even must consist of more than only two components. A Stoic disjunction actually has the form "ἤτοι ... ἤ ... ἤ ... / either ... or ... or ...". How many elements are to be listed varies from case to case, depending on various factors. If the disjunction is formed from a proposition and its negation, there is no space for more than two components, otherwise there often is. The crucial point is that not all of these components are false, but one is true, and that they exclude each other, so that by accepting one of them as the true one the rest are ruled out. Hence, the Stoic disjunction is exclusive. But in order to be a valid disjunction, it has to be not only exclusive but, according to what Gellius reports,[27] also exhaustive – version A –; or at least it must include the one true item – version B –. The Stoics's day/night example serves not only as a standard example for a disjunction in general but in particular also for a disjunction that fulfills the additional requirement.

In the available sources there are several Stoic standard examples preserved that instantiate tripartite disjunctions, mostly furnished with a note indicating that the disjunction would be false if because of the omission of one of its elements it was incomplete, or not exhaustive enough: "Pleasure is either good or evil, or it is neither good nor evil" (Gellius, NA 16. 8. 12 = FDS 976). The second example applies the same three predicates to health, and a third one to the orders given by a father to his son.[28] A further example certainly goes back to Chrysippus; it tells the story of a dog hunting a game animal. When coming to a crossroads, in order to decide which way to follow, the dog forms a disjunction and applies a syllogism: "The animal fled either this way or that way or the third way; now not the first and not the second; therefore the third" – and that is the direction into which the dog runs without further sniffing (Sextus Empiricus, PH 1. 69 = FDS 1154). Finally, the example of sitting and walking, brought up by Proculus, occurs also in Galen, but is enlarged there a little, so that a disjunction results that is composed of five parts; presumably it was considered to be correct (Galen, IL 5. 2 = FDS 977).[29] Today, an example still more

---

[26]Cf. esp. [8, p. 115 f.]; also [9, p. 93 ff]; and [4, p. 104, 5, p. 92 f.].

[27]See the quotation in the next section.

[28]Sextus Empiricus, PH 2. 150 = FDS 1111, AM 8. 434 = FDS 1110; Gellius, NA 2. 7. 21 f. = FDS 975).

[29]It should be noticed that this disjunction, though presumably correct, could hardly be considered exhaustive. For, what about dancing, or standing on one's head, or further positions that are shown

complex would be the Sudoku game in which the digits 1 to 9 form an exhaustive disjunction containing nine elements.

If we compare the examples with the two afore-mentioned versions to describe the disjunction, it appears that they do not really correspond to Gellius's report, but rather back up version B. While the report seems to postulate that all the possible components must always be mentioned irrespective of the circumstances, the examples do not demonstrate precisely this. There is no exhaustiveness check included, and they demonstrate only how important it is not to omit a component that might prove to be the true one. In particular the first examples, if with them the neither/nor-case was omitted and the alternative referred simply to good and evil, the disjunction would be false, especially according to the Stoics, since in antiquity just these philosophers considered pleasure and health to be neither good nor evil. In the light of the examples, then, to be valid a disjunction need not be actually exhaustive; instead, it has to comprise those partial propositions that are candidates for being true, but it may omit components that are certainly false; yet the rule reported by Gellius seems to insist on these propositions, too. To this extent, the ambiguity between version A and version B, observed above, is confirmed by the examples, not solved by them; and from a different angle we will meet it also in the next section.

Irrespective of this difficulty, then, a proper Stoic disjunction is a non-simple proposition composed of several partial propositions that are connected by the conjunction "or"; these components exclude each other, precisely one of them is true, and more or less they exhaust the matter they concern. However, to supplement this description by a suitable definition is not as easy as it appears to be in Proculus. The ambiguity commented upon constitutes one problem. Further difficulties for a proper definition of the disjunction arise in various ways. For instance, "proposition" is a rather loose translation for the Stoic term $\dot{\alpha}\xi\acute{\iota}\omega\mu\alpha$; "statable" would be more precise. This term, however, evokes the Stoic theory of *lekta* and the question how to interpret the conjunctions in terms of this theory, in particular the conjunction "or" and the syntactical rule mentioned above, a question that demands a proper understanding of the mutual exclusion between the components of a disjunction.[30] This topic also causes difficulties since as a matter of principle a disjunction may include more than only two components.

In taking up this line of thought, the first requirement for a definition of the disjunction will be that, corresponding to the conceptual conditions at hand, precisely one of the dis-connected propositions is true, and consequentially (at least) one is false. As soon, then, as the truth value of one component is certain, it is also certain that all the other components are false. Yet, to conclude conversely from the falsity of one component to the truth of the other, as Proculus says, is possible only in the case of bipartite disjunctions, not in all cases. What is more, to define

---

by gymnasts or performers, but not listed so far? So, the example seems to support the weaker requirement only.

[30]For an approach of this kind see [2].

the general case is more complicated, and how the Stoics tried to do that, has been unsatisfactorily passed down to us. Their definition of the disjunction, therefore, is rather to be reconstructed from the testimonies, than taken as properly documented.

While Proculus in his efforts to characterize the disjunction uses modalities, according to Galen, Gellius and Sextus Empiricus the Stoics apparently used the incompatibility concept, similarly to the position taken by Chrysippus in the case of the conditional. Gellius, for instance, writes[31]:

> Omnia autem, quae disiunguntur, pugnantia esse inter sese oportet, eorumque opposita, quae αντικείμενα Graeci dicunt, ea quoque ipsa inter se adversa esse. Ex omnibus, quae disiunguntur, unum esse verum debet, falsa cetera.
>
> But all the disjuncts must be in conflict with each other and their contradictories (which the Greeks call αντικείμενα) must be contrary to each other, too. Of all the disjuncts one must be true, the remaining ones false.[32]

According to this, combined with what Galen explains about complete and incomplete incompatibility (Galen, IL 5. 1 f. = FDS 977), the Stoics postulated a mutual incompatibility of the partial propositions. This incompatibility constituted so-called incomplete incompatibility, i. e. it ensured that at the most one of the components can be true; from the truth of one of them the falsity of the others follows. In order to achieve so-called complete incompatibility, too, it must be ensured by a second postulate that the components, in addition to impossibly being *true* together, could not be *false* together either. In the case of a bipartite disjunction this condition is fulfilled if not only the two partial propositions but also their negations are mutually incompatible; in just this form the criterion is to be found in the sources. But in the case of more than two components, the negations of at least two of them must be mutually compatible. In this case, therefore, we cannot request that the negation of each part of the disjunction is incompatible with the negation of each other part; but in addition to the first incompatibility we can, as M. Frede put it, postulate merely that the negation of each part of the disjunction is incompatible with the conjunction of the negations of all the other parts.[33]

Before passing on from the disjunction to the subdisjunction(s), three things should be noticed.

(1) Though we have no further information how the Stoics explained their incompatibility concept, it is reasonable to assume that they elucidated it, and did not merely go back to the foregoing requirement by saying that incompatible propositions cannot be true together. For not only the modern disjunction and the material implication are interrelated. The same holds true also for the Stoic disjunction and Chrysippus's concept of the true conditional, both of which

---

[31] Gellius, NA 16. 8. 13 = FDS 976. The passage is immediately followed by the passage quoted in the next section.

[32] English translation by S. Bobzien [4, p. 110, 5, p. 96].

[33] "... daß die Negation eines jeden Disjunktionsgliedes mit der Konjunktion der Negationen der übrigen Disjunktionsglieder unverträglich ist, nicht aber, daß die Negation jedes Disjunktionsgliedes mit der Negation jedes anderen Disjunktionsgliedes unverträglich ist" [9, p. 96].

refer to incompatibility. Hence, it is only if they had something more to say on incompatibility that they could – as far as the conditional is concerned – claim that Chrysippus's concept essentially differed from the proposals developed by Philo and Diodorus Cronus.[34] So, from the discussion concerning Chrysippus's concept of the conditional, there may be expected further hints with regard to the disjunction. But, as long as we have no convincing idea how the Stoics wanted to clarify the incompatibility concept, we are not in a position to offer a satisfying reconstruction of their concept of disjunction.

(2) Likewise it is not entirely clear how the truth-functional description of the disjunction – I called it "the first requirement" – and the explanation in terms of incompatibility are interrelated. S. Bobzien, however, reads the Stoic definition of the disjunction explicitly as a definition in two steps: after starting with the truth-functional criterion it uses the incompatibilities for tightening the initial criterion [4, p. 109–111, 5, p. 96 f.].

(3) Proculus's use of modalities in describing the disjunction should be understood as being absolutely equivalent to the more common references to incompatibilities. To interpret it in this way corresponds to the advice formulated at the end of Sect. 2.3.1; it is backed up by Galen's mode of expression in the passage just mentioned; and finally, we would not have sufficient reasons for pointing out major differences here. On the contrary, by introducing new aspects here Proculus would have been deviating from his Stoic forerunners, and would have weakened or even lost the dialectical justification he needed for his juridical purposes.

2.3.4   The Subdisjunction in General

When coming from the disjunction, we may observe first that all sources available for us agree on the view that the step toward the subdisjunction is taken by weakening the requirements, though which requirements are the crucial ones here is less clear. Is it the mutual exclusion, or the (more or less complete) exhaustiveness, or both of them? As long as this is under discussion and as the respective weakenings are not specified, difficulties concerning an appropriate definition of subdisjunctions are only natural. This might explain why only very few and late ancient texts explicitly testify Stoic discussions about "or"s different from the disjunctive (exclusive and exhaustive) "or", and why these passages differ in crucial respects.

---

[34]On the Chrysippean concept of the conditional see esp. FDS 947–965. As to the interpretation M. Nasti De Vincentis [22] and [23], started a new approach. Though it is ambitious and challenging, with regard to incompatibility and disjunction it proves very problematic; see Castagnoli's review [7, p. 185 f.].

1. As to the three weakenings mentioned, it seems that each and every one of them suffices to turn disjunctions into subdisjunctions. Gellius, however, writes with regard to the components of a disjunction (Gellius, NA 16. 8. 14 = FDS 976):

> Quod si aut nihil omnium verum aut omnia plurave, quam unum, vera erunt aut quae disiuncta sunt, non pugnabunt aut quae opposita eorum sunt, contraria inter sese non erunt, tunc id disiunctum mendacium est et appellatur παραδιεζευγμένον.
>
> For, if none at all of them is true, or if all, or more than one, are true, or if the contrasted things are not at odds, or if those which are opposed to each other are not contrary, then that is a false disjunctive proposition and is called παραδιεζευγμένον (subdisjunctive proposition).[35]

According to this passage, it doesn't matter which one of the various weakenings under discussion is performed; in any case a subdisjunction results.

2. Next, considering the weakenings separately, we may state regarding the mutual-exclusion condition that all the testimonies we have confirm that the weakening of this condition results in a subdisjunction; or to be more precise: it results in a subdisjunction at least when it means that not *precisely* one but *at least* one of the partial propositions is true, i. e. if the "or" connection merely indicates that not all of the partial propositions are false. Apparently this type even became the standard form of the subdisjunction. For Galen and Apollonius Dyscolus as well as in the *Epimerismi Homerici* passage it seems to be *the* subdisjunction (see FDS 977, 979 f., 1153). And even for Proculus it is just that kind of subdisjunction he was actually interested in.

3. The other requirement, namely that a disjunction must be (more or less) exhaustive, is stressed in many places in order to make sure that the disjunction contains at least one true constituent. From the Stoic examples for tripartite disjunctions mentioned above, the examples regarding pleasure, health, and the orders given by a father to his son are of this kind. As soon as the neither/nor possibility in them is omitted, the disjunction turns out to be not exhaustive (or not exhaustive enough), and for just this reason it is false. Yet there are other cases of not-exhaustiveness, quite different in character, that render the *prima facie* disjunction an unexpected kind of subdisjunction.

Let's take one of the Stoic examples of a disjunction composed of more than merely two parts, for instance the example of sitting and walking etc., enlarged in Galen to a disjunction composed of five parts; though surely it could be further enlarged, presumably it was considered to be correct: "Either A is sitting, or A is standing, or A is running, or A is walking, or A is lying down". Let's assume, now, that the first as well as the second of the components is false. We may conclude then: "or the third or the fourth or the fifth", and modify this (linguistically questionable) intermediary result into "Either the third or the fourth or the fifth". This is surely a true non-simple proposition, since it is composed of several simple propositions, and since it is a logically correct consequence from true premises and thus must be true as well. But is it still a

---

[35]English translation by Rolfe [24].

disjunction or just a subdisjunction? On the one hand it still includes precisely one true component. But on the other hand it is exhaustive only because of the context and not in itself. For this very reason the ambiguity pointed out above, the ambiguity between version A and version B of the disjunction concept, reappears here. The proposition under discussion does not really meet the stronger version A, in other words Gellius's definition of the disjunction, but might be considered a disjunction that is exhaustive only because of its context. Yet it could also be seen as a new kind of subdisjunction; at least it fulfils the criterion for subdisjunctions quoted above from Gellius.

This kind of problematic case unavoidably arises as soon as Chrysippus establishes his so-called indemonstrable arguments. So there must have been a very old discussion about this, though the testimonies do not mention it at all. Maybe the term "παρα-διεζευγμένον / sub-disjunction" could be explained best by reference to the unexpected type of subdisjunction circumscribed so far; and perhaps this type was even the oldest one discussed by the Stoics. Later, however, it was not discussed any more. But in the present context it must be emphasized because of the disjunction concept, and in order to show that, depending on this concept, a subdisjunction could possibly be constituted not only by weakening the requirement of mutual exclusion but also by separately weakening the requirement of exhaustiveness in the strong sense of version A; that would constitute a different type of subdisjunction.

4. To weaken the two requirements jointly is also possible, of course. And in the available sources there is at least one passage that might be read as an example for this third kind of subdisjunction. When instantiating syllogistic conclusions from subdisjunctions, Galen forms the following example: "The distribution of nourishment from the belly to the whole body occurs, either by the food being carried along of its own motion, or by being digested by the stomach, or by being attracted by the parts of the body, or by being conducted by the veins" (Galen, IL 15. 1 = FDS 1153; translated in [14]). In view of this list of possibilities, every ancient physician would have agreed that more than only one partial proposition may be true; but additionally it would be difficult to consider the series of components exhaustive. To this extent, *both* requirements for correct disjunctions are weakened here.

2.3.5 Proculus's Differentiations

Proculus opens his remarks by pointing out that the words "ille aut ille" ("So-and-so or So-and-so") occur in disjunctive as well as in subdisjunctive speech; a little later he adds that disjunctive and subdisjunctive propositions are syntactically similar ("simili figuratione"); and at the end of his differentiations he gives examples completely conforming to this, since they differ from disjunctions neither lexically nor syntactically. Accordingly, in order to discriminate between disjunctions and subdisjunctions (safely), it is not enough to check linguistic phenomena, but one has to think about the context too. The same applies to the two kinds of subdisjunctions; they, also, can be identified only by taking notice of the context.

Of course, this has to do something with the fact that in Latin it would be extremely difficult to establish a convention parallel to the Stoic preference for "ἤτοι ... ἤ ...". Moreover, it corresponds to linguistic usage, and to the juristic practice that treats only certain cases, precisely enumerated in Justinian's codex, as special cases, i. e. as cases that are discriminable not on linguistic grounds. Besides all this, however, Proculus hints at the insufficiency of purely linguistic means for the discrimination of disjunctions and subdisjunctions, and his more or less explicit pointing to the actual use and the context of those forms is also remarkable from a logical point of view. In this respect it indicates that the famous Stoic formalism does not work in the case of "or". In order to specify the type of "or" in special cases, it is in fact unavoidable to reflect on the circumstances and to determine the sense of what is said.

We will come back to this topic when reflecting on Proculus's response to the ambiguity outlined above.

In view of the various possibilities for turning disjunctions into subdisjunctions, it is quite obvious that Proculus determines his subdisjunctions solely by weakening the requirement of mutual exclusion, and he weakens it in two ways so that two kinds of subdisjunctions are defined: instead of requiring that one and only one of the components is true, the condition is merely that either not all of them are true or not all of them are false. To put it otherwise, the new requirement is, either that at least one of the partial propositions is false (but possibly all of them), or that at least one of them is true (but possibly all of them). The former variant constitutes Proculus's subdisjunction type 1, nowadays known as the exclusion and as Sheffer's stroke, while the latter variant forms subdisjunction type 2, and later became the standard or main subdisjunction, or even the only subdisjunction, and at present it is known as the disjunction in the modern sense of the term. In antiquity, it constituted the crucial precondition with regard to the so-called laws of De Morgan, the first absolutely unambiguous application of which may be observed in a passage from Julianus Salvius, a Roman lawyer of the second century AD.[36]

All this is obvious. Less obvious is how to assess the fact that the former kind ("Sheffer's stroke") was explicitly described solely by Proculus. About 100 years later it belonged to the scope of Gellius's implementation of subdisjunctions,[37] while neither Galen nor Apollonius Dyscolus left any room for it since for them Proculus's second kind constituted subdisjunctions. Afterwards, it disappeared completely and reappeared only in modern propositional logic. Apparently, even the symmetry of Proculus's account and Gellius's definition could not protect this kind of subdisjunction. From the second or third century onwards no one seemed to need

---

[36]D. 34. 5. 13. 6 = FDS 978a. If not really the first, the passage is at least one of the very first applications of De Morgan's laws.

[37]See the passage from Gellius quoted above.

this kind of subdisjunction anymore,[38] so that Proculus's type 2 turned out to be the only kind. This will have been one of the more significant preconditions for the later – the more modern – development that rendered the remaining *sub*-disjunction the new disjunction we have in contemporary logic.

In drawing his distinctions Proculus relied heavily on Stoic logicians, and the credit for his subtleties belongs more to them than to him, the more so as Proculus himself had no need for the type 1 subdisjunction in itself but only used it as an intermediate step in pointing out the other subdisjunction. The Stoic approach to subdisjunctions presumably resembled more Gellius's than Proculus's account, since Gellius, though younger, reports a wide variety of reasons that lead from disjunctions to subdisjunctions while Proculus gives a rather short description. However, this comparison confirms the thesis that in reporting Stoic doctrine Proculus is at the same time filtering and putting aside certain aspects of it.

To approach his subdisjunctions from the disjunction, Proculus weakened the requirement of mutual exclusion. But what did he do with the second criterion, the requirement of exhaustiveness? Though not mentioning it explicitly, he did not set it aside completely – for two reasons. First, instead of *eliminating* the question of completeness, the juridical context renders it a *trivial* problem, since persons who are not mentioned in testamentary dispositions (normally) are ruled out from the beginning; these dispositions refer to all and only all persons named or described in the respective documents. Second, when coming to his kinds of subdisjunction Proculus twice uses the expression "ex propositis finibus" ("because of the presupposed limitations"). This expression does not *specify* any requirements but *presupposes* that the appropriate limitations are already fixed, i. e. determined by the context in which the subdisjunctions are used. Proculus, therefore, does not eliminate the aspect of exhaustiveness but contextualizes it.

And this does not only apply to the subdisjunctions but also to the disjunction. For only then is it justified to assert explicitly – as Proculus does – that "subdisiunc-tivi ... genera sunt duo" ("there are two kinds of subdisjunctive [proposition]"), i. e. that there are no further kinds but solely the two characterised by Proculus in the frame of weakening the mutual-exclusion requirement. Hence, also in the disjunction the aspect of exhaustiveness is presupposed as being determined by the context in which the disjunction is used. This view entitles Proculus as well as his Stoic referees to state that there are merely two kinds of subdisjunctions.

At the same time it clarifies the ambiguities described above, and transcends the idea that linguistic standardisations could suffice to determine disjuntions safely. (a) The ambiguity concerning the concept of disjunction is decided in the sense of version B. It is clarified by the assumption that the condition of exhaustiveness is considered to be fulfilled as long as it is guaranteed that one of the disjunction's

---

[38]To this day, even in jurisprudence it is difficult to find a relevant example for this kind of "or". Bocheński and Menne [6, p. 24] substituted Proculus's example and pointed to nationalities, at least nationalities in interpretations in which a person might be either English or German, i. e. not both, though possibly neither of them. But one wonders whether this is a typical juristic example.

components is true. And thereby the third kind of subdisjunction that was taken into consideration in the preceding section immediately disappears. Hence, Proculus does not only report traditional Stoic logic and filter it with regard to his juridical purposes. He also modifies it slightly by disambiguating the exhaustiveness criterion for a correct disjunction. (b) Cicero, Galen, and other ancient authors blamed the Stoics for trying to standardise or even to regiment human language; the linguistic surface should immediately show what an expression means (see e. g. FDS 232–242.). The Stoic rules for disjunctions and other non-simple propositions[39] might be seen as justifying this criticism. Proculus, however, says that in the case of disjunctions and subdisjunctions the linguistic form does not differ. At least here, then, it will not work to make the linguistic form of an expression unambiguously show its logical form. Thus, the aforementioned ideal of linguistic standardisation is transcended. It will always remain crucial to find out the sense of what is said.

Maybe the credit for this belongs not so much to him but to his immediate Stoic forerunners. In this case the modification is not so much his but their work. This would mean that there were two groups of Stoic logicians. Both of them discussed the disjunction concept but disagreed on how to interpret the requirement that a valid disjunction must include one true component. One group specified it in the sense of version A, and developed a definition similar to what is later reported by Gellius, while the second group favoured version B and showed up in Proculus's *Letters* only. The bifurcation between the two groups, then, will have taken place during the first century BC. Among other things the Stoics were known for having a major diversity of opinion in their school.[40] The case of the (sub)disjunction could serve as an interesting example of this diversity.

2.4   Conclusion

In Rome there was a strong juridical demand for the entitlement to interpret the conjunction "or" in certain contexts in a non-exclusive sense. Proculus satisfied this demand by going back to the logical textbooks of the Stoics. There he found appropriate differentiations and pointed out a non-exclusive "or" that was well known to (or at least convincingly discussed by) Stoic logicians and that did not differ from the exclusive "or" on the linguistic level. In explaining his finding he put various aspects of the Stoic reflections on "or" aside, and clarified an ambiguity relating to the requirement that a correct disjunction must include precisely one true component. According to him, for the correctness of a disjunction it was sufficient that the circumstances under which the disjunction is used already guarantee that one of the disjuncted components is true. As a consequence there was room for only two subdisjunctions, and one of them was just the "or" needed for the juridical aims in question.

---

[39] See Sect. 2.3.3 at the beginning.

[40] See FDS 225–226, and, for instance, the famous Stoic explications of the human *telos*.

After Proculus the other subdisjunction was mentioned only by Gellius and by no one else. Apparently, it was considered practically irrelevant and disappeared. The practically relevant subdisjunction became the main and even the only subdisjunction, differing from the disjunction in one respect only: while the claim expressed through the subdisjunctive "or" was that *at least* one component is true, the one expressed through the disjunctive "or" was that *precisely* one partial proposition is true.

In the subsequent developments, on the one hand a tendency evolved to minimize even the difference between subdisjunctions and conjunctions. This development can be already observed in Apollonius Dyscolus and in the passage from the *Epimerismi Homerici* (cf. FDS 779f.); it is apparent in the passage from the *Codex Iustinianus* quoted above in Sect. 2.2.3, and it has been documented convincingly by M. Frede (see [9, p. 99f.]). But, on the other hand, there was also a serious interest in keeping up the difference between "and" and "or", especially in juridical contexts. For instance, in the first half of the third century the lawyer Iulius Paulus stressed this difference in various places by saying that in cases of conjunctively given conditions, all of them must be observed, while in cases of disjunctive formulations merely one is to be fulfilled (D. 28. 7. 5; 50. 16. 28; 50. 16. 29; 50. 16. 53). Similar remarks are preserved from the lawyers Modestius and Papinianus (D. 35. 1. 51 and 36. 2. 25, respectively). In all of these passages the term "disjunction" covered both Proculus's disjunction and his second subdisjunction.

In these developments, the term "subdisjunction" was seemingly more and more intended to cope with the similarities between "or" and "and", while the term "disjunction" rather had to cope with the differences between them. Thereupon, regarding disjunction, one next step offered itself, namely to use the weaker criterion for the leading perspective, to make the remaining Stoic and Proculian subdisjunction the new disjunction and to rename the old disjunction. This step was taken in modern times. But inasmuch as it was prepared for by Roman lawyers and in particular by Proculus, we could say that jurisprudence not only received something from logic, it also gave something back – a rather rare event.

References

1. M. Armgardt, Zur Bedingungsdogmatik im klassischen römischen Recht und zu ihren Grundlagen in der stoischen Logik. Tijdschrift vor Rechtsgeschiedenis **76**, 219–235 (2008)
2. J. Barnes, What is a disjunction? in *Language and Learning*, ed. by D. Frede, B. Inwood (Cambridge University Press, Cambridge, 2005), pp. 274–298. (Repr. in: J. Barnes, *Logical Matters. Essays in Ancient Philosophy*, ed. by M. Bonelli (Clarendon Press, Oxford, 2012), pp. 512–537
3. R.A. Bauman, *Lawyers and Politics in the Early Roman Empire. A Study of Relations Between the Roman Jurists and the Emperors from Augustus to Hadrian* (C. H. Beck, München, 1989)
4. S. Bobzien, Logic/The stoics, in *The Cambridge History of Hellenistic Philosophy*, ed. by K. Algra, J. Barnes, J. Mansfeld, M. Schofield (Cambridge University Press, Cambridge, 1999), pp. 92–157

5. S. Bobzien, Logic, in *The Cambridge Companion to The Stoics*, ed. by B. Inwood (Cambridge University Press, Cambridge, 2003), pp. 85–123

6. I.M. Bocheński, A. Menne, *Grundriss der Logistik* (Ferdinand Schöningh, Paderborn, 1954)

7. L. Castagnoli, Review of Mauro Nasti De Vincentis, *Logiche della connessività* (Bern, Stuttgart, Wien: Haupt 2002). Elenchos **25**, 179–192 (2004)

8. T. Ebert, *Dialektiker und frühe Stoiker bei Sextus Empiricus. Untersuchungen zur Entstehung der Aussagenlogik* (Vandenhoeck & Ruprecht, Göttingen, 1991)

9. M. Frede, *Die stoische Logik* (Vandenhoeck & Ruprecht, Göttingen, 1974)

10. A.M. Honoré, Proculus. Tijdschrift vor Rechtsgeschiedenis **30**, 472–509 (1962/1963)

11. K. Hülser, *Die Fragmente zur Dialektik der Stoiker. Neue Sammlung der Texte mit deutscher Übersetzung und Kommentaren*, 4 vols. (Frommann-Holzboog, Stuttgart, 1987/1988) (abbr.: FDS)

12. P.E. Huschke, *Iurisprudentiae anteiustinianae quae supersunt, in usum maxime academicum compositae*, 5th edn. (B. G. Teubner, Leipzig, 1885) (6th ed. 1908 by E. Seckel & B. Kuebler)

13. J.C. Joerden, *Logik im Recht. Grundlagen und Anwendungsbeispiele* (Springer, Heidelberg, 2005), 2nd ed. 2010

14. J.S. Kieffer, *Galen's* Institutio Logica. *English Translation, Introduction, and Commentary* (The Johns Hopkins Press, Baltimore/Maryland, 1964)

15. C. Krampe, *Proculi Epistulae. Eine frühklassische Juristenschrift* (C. F. Müllser, Karlsruhe, 1970)

16. P. Krüger, T. Mommsen, *Corpus iuris civilis. Institutiones, recognovit Paulus Krueger; Digesta, recognovit Theodorus Mommsen* (Weidmann, Berlin, 1877) (editio stereotypica 5: 1889; 6: 1893; 13: 1920; 16: 1954)

17. Y. Lassard, A. Koptev (eds.), *The Roman Law Library*. http://droitromain.upmf-grenoble.fr/

18. O. Lenel, *Palingenesia Iuris Civilis,* vols. I, II (Officina Bernhardi Tauchnitz, Leipzig, 1889)

19. D. Liebs, Rechtsschulen und Rechtsunterricht im Prinzipat, in *ANRW*, vol. II. 15: *Prinzipat*, ed. by H. Temporini (De Gruyter, Berlin/New York, 1976), pp. 197–286

20. L. Marrone, Le *Questioni Logiche* di Crisippo (*PHerc. 307*). Cronache Ercolanesi **27**, 83–100 (1997)

21. J. Miquel, Stoische Logik und römische Jurisprudenz. Zeitschrift der Savigny-Stiftung für Rechtsgeschichte: Romanistische Abteilung **87**, 85–122 (1970)

22. M. Nasti De Vincentis, *Logiche della connessività. Fra logica moderna e storia della logica antica* (Haupt, Bern/Stuttgart/Wien, 2002)

23. M. Nasti De Vincentis, Conflict and connectedness: Between modern logic and history of ancient logic, in *Logic and Philosophy in Italy. Some Trends and Perspectives. Essays in Honor of Corrado Mangione on His 75th Birthday*, ed. by E. Ballo, M. Franchella (Polimetrica, Monza, 2006), pp. 229–251

24. J.C. Rolfe (transl.). *The Attic Nights of Aulus Gellius, in three volumes,* vol. 3. The Loeb Classical Library vol. 212 (Harvard University Press, Cambridge MA, 1927), revised and reprinted 1952, 1961, 1967

25. F. Schulz, *Geschichte der römischen Rechtswissenschaft* (Hermann Böhlaus Nachfolger, Weimar, 1961) (= *History of Roman Legal Science*. At the Clarendon Press, Oxford, 1946)

Chapter 3
Disjunctive Statements in Roman Legal Arguments

Two Examples by Julian and Quintilian

Markus Winkler

Abstract So far, modern scholars of Roman law interested in the methodology employed by individual jurists have treated logic with different emphasis. Taking a bottom-up approach, one can search for specific texts containing explicit references to logical instruments. Their interpretation may benefit from logical concepts. Taking a top-down approach, one may look for different ways to integrate logic with the traditional picture of Roman casuistry. This contribution intends to exemplify both approaches. First, it illustrates how the Roman jurist Julian deals with an explicit disjunctive statement lying at the heart of a simple case where a debtor made an erroneous performance under an obligation. Secondly, a link is proposed to Roman procedural formulae, interpreted as conditional statements. Given Julian's bare text, various missing elements from Roman legal practice have to be added to establish this link. A second example taken from Quintilian's Lesser Declamations shows how these missing elements may have looked like in actual practice. Incidentally, the second text offers an illustration for an alternative, implicit use of disjunctive statements in legal argumentation.

3.1 Introduction

3.1.1 Greek Influence

Contemporary scholars of Roman law have been looking at the subject of logic from different angles. One line of approach aimed at a general study of the influence of Greek science and philosophy on the development of classical Roman jurisprudence – an influence commonly seen as active since the final decades of the Republic. Opinions range from the idea of a veritable revolution of Roman legal thinking to more moderate stances limiting the influence of Greek science to

M. Winkler (✉)
Rechtswissenschaftliches Institut, Universität Zürich, Zürich, Switzerland
e-mail: markus.winkler@uzu.ch

© Springer International Publishing Switzerland 2015

M. Armgardt et al. (eds.), *Past and Present Interactions in Legal Reasoning and Logic*, Logic, Argumentation & Reasoning 7, DOI 10.1007/978-3-319-16021-4_3

isolated inroads affecting certain methods of an otherwise already well established legal practice.[1] Berman aptly set the tone for the sceptics when he summarily denied that Greek dialectical reasoning resulted in "the intermarriage of Roman law with Greek philosophy" [4, p. 135] an event that came about only some thousand years later when the newly established European universities embarked on their scientific study of law.

Another line of approach led scholars to adopt a more limited perspective, focusing on the methods employed by selected individual jurists.[2] This type of study may suffer from the view that Roman jurisprudence was shaped less by individuals than by collective efforts of recognized jurists over longer periods of time.[3] While being conscious of the looming danger stemming from suggestive but undue generalizations of isolated cases, I tend to value the advantage in staying focused and working closely on a manageable selection of transmitted texts from the Digest.

A third line of thought uses techniques like rhetoric or logic in an effort to interpret the writings of Roman jurists, gaining additional insights through an interdisciplinary approach. Miquel's 1970 article [20] on the influence of Stoic logic on Roman jurists is an inspiring early example.[4] Employing slightly "modernized" techniques of assertoric logic, which nevertheless have their historical roots in antique sources like Proculus' description of disjunctive statements transmitted in D. 50,16,124,[5] he suggested – among other points – various emendations to Julian's *liber singularis de ambiguitatibus* before summarizing his results, claiming that Roman jurists were able to "think axiomatically".[6]

This contribution shares aspects of the second and third line of approach. Looking at one particular logical device, it modestly aims at illustrating practical applications of disjunctive statements as characterized in D. 50,16,124. The primary example is one of several of its kind, selected from Julian's digest.[7] Less modestly, I propose to set logic as a tool into perspective with the traditional Roman casuistry, for which I am briefly sketching three functions for logic in law. The sketch will also highlight the probable limits of applying logic to law. A second example taken from Quintilian's *Lesser Declamations* shall provide additional details missing in Julian's bare text to illustrate this proposal.

---

[1]The origin of this interest can be seen in [30]. For a summary overview on the literature since then, see [13, p. 286] and [25, p. 1], in particular [26].

[2]See for example [8, 17, 32]. A general perspective on the theme can be found in [11].

[3]See [25, p. 4] and [10].

[4]See also Waldstein [31], who saw a considerable potential for the use of logic in law.

[5]See K. Hülser's contribution to this volume.

[6]Various authors have since demonstrated that there is no real need for proposing emendations [1, 9, 33, 34, 35].

[7]I am proposing a broader and more stringent investigation on logic and elementary mathematics as techniques in Julian's digest in my dissertation at the University of Zurich.

3.1.2 Logic in Perspective

3.1.2.1 Outlining Its Scope in Antiquity

For the purposes of this study, I am looking at logic from the perspective of dialectic as it was conceived in Antiquity.[8] Aristotle's *Organon* serves as a reasonably sound guide to the content of logic in the appropriate historical setting, to which those few glimpses of Stoic thought have to be added, that transpire to us through writings of authors such as Cicero, Sextus Empiricus or Boethius. Two broad streams can be distinguished: assertoric or non-modal logic and modal logic. Assertoric logic includes the foundations of what today is known as propositional calculus and predicate logic, albeit without the formal apparatus introduced by Leibniz and later in the nineteenth and twentieth centuries by logicians like Wittgenstein or Frege. Assertoric logic looks at propositions, their truth-values[9] and combinations of propositions, formed by using the connectives of conjunction, disjunction and the conditional.[10] The Stoics widely discussed logical connectives in their days (see [12, p. 159]). By contrast, Aristotle's syllogistic logic looked at terms rather than at entire propositions. In modal logic, propositions are qualified in terms of the *modi* necessity, possibility and contingency. Modal logic is more powerful and proves better suited to questions of practical relevance, as it naturally supports the consideration of elapsing time. Both streams include detailed technical rules for logical inference that lead from a given set of premises to a conclusion.[11]

Looking at objects from this outline of antique logic arguably ensures that no concepts rooted in modern times interfere with a historically appropriate study of the texts. An occasional recourse to modern-day formalizations may nevertheless prove useful in approaching those texts. Certainly, Wittgenstein introduced modern-style truth-value tables only around 1921. There is nothing to be said, in my view, against their use in analyzing antique cases as Miquel did in his 1970 paper, as long as we relate back any results so derived to the text analyzed in its historical context.

---

[8]For a precise presentation of the not always one-to-one relationship between logic and dialectic, see the overview in [6] and [12, p. 139].

[9]The concept as such was known to Aristotle, see e.g. *Metaphysics* V, 29 (1024b).

[10]Whether negation can be looked at as a connective in Stoic parlance is a debatable point. For more background see [12, p. 160]. In fact, the key prerequisite for any logical analysis of legal texts is the appropriate identification of the relevant truth-bearers. For the Roman period, this relates to the question whether today's idea of a "proposition" is equivalent to the Stoic concept of *axioma*. This topic shall not be further discussed here. Suffice to say that *axiomata* seem reasonably close to propositions to justify such analysis. See [12, p. 139] or [2, p. 45].

[11]For Aristotle's logic see e.g. [23]; for Stoic logic see [5].

3.1.2.2 Three Functions for Logic in Law

Looking at D. 50,16,124, we may ask whether there is any evidence for practical applications of disjunctive statements in Roman legal writings. Julian's *liber singularis de ambiguitatibus* appears to be less an example for such an application than a theoretical reflection on the ambiguity inherent in the disjunction.[12] If this characterization is correct, then the *liber singularis* represents a rare instance of a contemporary treaty on legal methodology. Bund defined the concept of legal method as the path leading to the right decision.[13] Hypothesizing that logic was an integral part of legal methodology in Roman times seems chancy without having first looked at the more general question of how logic can integrate with the traditional picture of Roman casuistry. For this purpose I am going to sketch three possible functions for logic in law. The following comments are by no means limited to the specific subject of Roman law, as the advice to "think syllogistically" by an eminent member of the US Supreme Court nicely illustrates (cf. [24, p. 41]).

Logic comes in completely naturally through the conditional understood linguistically as an "if – then"-sentence. In the more abstract setting of sociological system theory ("Systemtheorie"), Luhmann [19, p. 60] proposed a binary code distinguishing between "right" and "wrong" as the heart of any legal system. Based on this code, so-called "programs" are responsible to steer people's behaviour by setting corresponding expectations [18, p. 432]. For Luhmann [19, p. 195], any program in a legal system can be described as a conditional program as opposed to a goal-based program ("Zweckprogramm"). As the outcome of a conditional program, a legal decision solely depends on a given set of explicit premises. Hence, conditional programs allow for coordination between the legal system and its environment. Any necessary facts can be identified cognitively from the environment. No other factors – like politics or economic objectives – are taken into account. Arguably, this concept represents an ideal view of the workings of a legal system. At the same time, it opens up a view on each of the three promised roles for logic in law. Firstly, as an instance of syllogistic inference, the "if – then" pattern is closely related to the notion of legal inference (see e.g. [14, p. 195]). Noerr [21, p. 76] has already pointed out that the conditional is one of the oldest forms of legislative language. In the next section, I will try to ascertain in more detail how close this relation may get in the Roman legal order. Secondly, logical techniques can be employed to prepare the necessary premises by analyzing the facts of a given case. The two examples taken from the writings by Julian and Quintilian will illustrate how logic can assist the lawyer in this task. Thirdly, logic might serve as a tool to ensure consistency across different decisions. This seems a most relevant function but several arguments speak against it playing an active role. Luhmann [19, p. 343] sees the only function of law in ensuring the quality of decisions through conditional programs. Consistency would have to be ensured by the very design of

---

[12]See footnote 6 for references to the relevant literature.

[13]See [8]: "Weg des Denkens, der zur gerechten Entscheidung führt".

all conditional programs in a legal system, a task many a modern legislator may find daunting. Such a legal system, were it to exist, would likely resemble a purely deductive, axiomatic system like Euclidian geometry.[14] By contrast, Roman jurists seem to have regarded their legal order – not to call it a system – more like a mosaic, punctuated by "islands of stability" [22, p. 41][15] provided by sporadic *leges* and *senatus consulta*. Whether Roman jurists were much concerned with consistency between decisions seems a debatable point.[16] Rather, they relied on the quality of their individual decisions based on a long established professional tradition. I will not consider this third function any further in this paper.

3.1.3   Highlighting the Conditional

The conditional shows a dual character as a logical connective and as an element of meta-logic. As a logical connective it moves on the same level as conjunction and disjunction. As logical inference it ascends to a more abstract level, controlling the relationship between the very objects of logical language.[17]

As a final step before turning to the examples, it seems indicated to look for possible traces of conditional programs in the Roman legal order. *Leges* and *senatus consulta* as elements of positive law are the most obvious instances for such programs. The second example by Quintilian will illustrate how the concept of a conditional program based on a given *lex* integrates with Roman legal practice. *Regulae iuris* could have played a largely similar role. However, there is still no little doubt about their binding force and application to individual cases.[18] In most cases of traditional Roman casuistry there is unfortunately neither a *lex* nor a *regula* to start with. The argumentative pattern of *casus, quaestio, responsio* apparent in Roman *digesta* offers at least a loose resemblance, as the example taken from Julian's digest will illustrate (see [25, p. 151]). The concept shows up much more stringently in Roman procedural *formulae* (see again [19. 197]). Some remarks of caution seem appropriate at this stage. Schmidlin [25, p. 153] pointed out that

---

[14]This ideal will be pursued by Spinoza with respect to ethics and by Wolff and Thomasius with respect to Jurisprudence.

[15]A legal system includes not only the relevant legal texts like statutes but also decisions and professional practice ("Rechtsbetrieb").

[16]When distinguishing *verum, utile* and *benignum*, Giaro [10, p. 427; 429] refers to solutions that are consistent with or in contradiction to the legal system. He goes on to describe legal fictions as turning into "pillars for the systematic coherence" of Roman law.

[17]Unlike the Stoics, who discussed the nature of the conditional at length, Aristotle's interest for the conditional seems to have been limited to this capacity, which is evidenced in Prior Analytic II, 2 (53b).

[18]See Giaro [10, p. 205] on the issue of normality of rules and maxims as well as the monograph by [25] and the critical comments by [21].

the Roman *actiones* did not form a comprehensive system.[19] They did not include precise sets of premises or conditions that could be matched up with the particulars of an individual case as modern lawyers do today when they sum up a case. Rather, the conditions are only indicated in general form and have to be further individualized, looking at the case at hand. It is indeed hardly to be expected that the Roman jurists thought along such abstract lines. As Stein [29, p. 48] aptly put it, the Roman jurists did not count the "concept of concept" among their mental equipment. Noerr [21, p. 78] adopts a more differentiated stance when he rejects the idea of a formal summing up as logical technique regularly employed by Roman jurists but admits that their approach can hardly be called by a different name whenever they based their reasoning on an authoritative text such as a *lex*. His view fully agrees, in my understanding, with the role of procedural *formulae* sketched above, when he stresses the importance of identifying the applicable *actio* and of gathering the necessary premises from the case. While such *formulae* may not form a comprehensive system in the modern sense, they nevertheless represent a patent recipe for analyzing an individual case along certain logical lines. The following text from Julian's digest underlines this thought (Iul. Pal. 128 = D. 44,2,24):

Si quis rem a non domino emerit, mox petente domino absolutus sit, deinde possessionem amiserit et a domino petierit, adversus exceptionem "si non eius sit res" replicatione hac adiuvabitur "at si res iudicata non sit".	When someone buys property from a non-owner, subsequently gets exonerated [from liability], the [true] owner having filed a suit against him, then loses possession of the property, and files a suit to recover it from the owner [who in the meantime has regained possession of the object], against the exception "if the property does not belong to him" he will find relief by the reply "and if the property has not been decided to be his".[20]

This is the classic case of a *rei vindicatio* to recover lost property as described in Gai. 4,16 and which can be found in the *Edict*. Lenel [16, p. 185] proposed the following formula for the action: *Iudex esto. Si paret rem qua de agitur ex iure Quiritium AA esse necque ea res restituetur, quanti ea res erit, ⟨exceptio⟩⟨replicatio⟩ tantam pecuniam iudex NN AA condemnato, si non paret absolvito.* The conditional form of the action becomes readily apparent.[21] The condition for deciding against the defendant is spelled out in the first part: "If it appears that the object in dispute belongs by Quiritian right to AA nor is it restituted". The legal consequence follows immediately: "NN is to be ordered to pay the corresponding value to AA" ("*NN condemnato*"). The additional part "*si non paret, absolvito*" makes clear that this condition is not only sufficient but also necessary to a decision against the defendant. In fact, the formula reflects the structure of a bi-conditional. The possible

---

[19]Giaro [10, p. 255] speaks of a "purely integrative" function of the Edict with respect to the civil law.

[20]If not specified otherwise, the translations are mine.

[21]The formulation is certainly no less explicit than modern versions like Art. 641 Abs. 1 ZGB or § 985 BGB.

interplay with exceptions and replications further sharpen this underlying logical structure. They are inserted into the formula ahead of the condemnation (marked with angular brackets in the text). The defendant NN claims an exception to the condemning condition by inserting the phrase "*si non eius [=NN] sit res*". The clause is set as the opposite of the defendant's claim that the property is actually his.[22] Likewise, the plaintiff inserts the phrase "*at si res iudicata non sit*".[23] This clause leads to condemning the defendant if the previous trial has not resulted in upholding the defendant's right to the title – which was the case in D. 44,2,24. The inclusion of exceptions and replications can lead to rather complex conditional programs. Their correct use requires at least a basic understanding of conditional statements. Strikingly, the interplay between opposing conditions is a recurring theme in Julian's writings.[24]

Eventually, the condition has to be individualized from the facts of the case: "*si paret*". It is most likely in the analysis of the facts of an actual case that the Roman jurist could have consciously turned to logical instruments.[25] This idea shall be spelled out in more detail in Julian's example in the next section.

3.2 Illustration

3.2.1 Julian

3.2.1.1 Iul. Pal. 161 (= D. 12,6,32 pr.)

pr. Cum is qui Pamphilum aut Stichum debet simul utrumque solverit, si, posteaquam utrumque solverit, aut uterque aut alter ex his desiit in rerum natura esse, nihil repetet: id enim remanebit in soluto quod superest.	Pr. When[26] anyone who is bound to deliver Pamphilus or Stichus, delivers both of them simultaneously, if after having delivered both of them, either both or one of them die,[27] he cannot recover anything; for what remains will be considered as performance under the obligation.

---

[22]See Gai. 4,119: *Omnes autem exceptiones in contrarium concipiuntur, quam adfirmat is cum quo agitur.*

[23]See Gai. 4,126: *Interdum evenit, ut exceptio, quae prima facie iusta videatur, inique noceat actori. quod cum accidat, alia adiectione opus est adiuvandi actoris gratia; quae adiectio replicatio vocatur [..].*

[24]See e.g. the complex of Iul. Pal. 520, 522, 464 und 465 on combinations of legacies with manumissions.

[25]In other words, the second function described above seems the most promising one to investigate.

[26]Assuming a temporal use, "*cum*" can be rendered as "when", "if", "while", "whenever". Scott translates by "where" which matches similar formulations used in the CFR. The use of "whenever" and to a lesser degree of "where" suggests a general rule, a meaning which is at least doubtful in the Latin original.

[27]Literally "cease to exist in nature".

Table 3.1 Inclusive and exclusive disjunction in Julian

A	B	A $)-($ B	A $\vee$ B	Comments
1	1	0	1	Debtor delivers both slaves: Why?
1	0	1	1	Debtor delivers Pamphilus only: Choice
0	1	1	1	Debtor delivers Stichus only: Choice
0	0	0	0	Debtor delivers nothing

This short text is taken from the title "*si certum petetur*" in the tenth book of Julian's Digest. There are numerous other texts in Julian's work with explicit or implicit references to logical instruments. The chosen text serves the purpose to illustrate the ideas presented in the general section without being unduly complex on the legal side.

Someone who owed Pamphilus or Stichus but delivered both slaves simultaneously may claim none back if one or both of them happens to die. The exact legal background of the case is not clear. Babusiaux [3, p. 385; 387] mentioned the case only briefly as an example for a choice debt in her work on Celsus's and Julian's writings on the edict "*si certum petetur*". The expression "*in rerum natura esse*" might also refer to a legacy. Especially, the verb "*solvere*" would match a *legatum per damnationem*, which could also be used as a *legatum alternativum*. In this case, the legatee has the right to choose between the alternatives stated in the will. The quoted text does not give any explanation why the debtor delivered both slaves, either. Interestingly, the exact legal background does not matter for the solution proposed by Julian.

From a formal point of view, the text resembles the reduced style of a *responsum* where the parts of *casus* and *quaestio* are reduced to a minimum and get commingled into a *cum*-sentence (see [25, p. 151; 156]). Accordingly, the reader must work out by himself both the essential facts and the underlying key question to get a full grasp on the case.

3.2.1.2 Formalization

The keyword "*aut*" explicitly refers to the use of the logical connective of disjunction. Proculus knew and described three different interpretations for the disjunction in the passage D. 50,16,124. Here, the debtor must correctly decide from the circumstances, which version is applicable to the obligation he entered into. In the modern form of a truth-value table, the two versions relevant in the present context are shown in Table 3.1.[28]

---

[28]The third version described by Proculus corresponds to the Sheffer-symbol A/B = "0111". It does not play a role in the arguments of Julian's Digest but appears as underlying structure in Quintilian's declamation Nr. 318.

Conceivably, the debtor was the victim of some ambiguity in the testamentary clauses, understanding *"aut"* as inclusive disjunction ($aut = \vee$) where the testator intended to use the exclusive disjunction ($aut =)-($). The debtor as heir charged with the legacy could interpret the testamentary clause to give him the choice to deliver both or either one or the other of the two slaves. Such an understanding would be extraordinary, to say the least. Following Julian's more lengthy recommendation at the end of his *liber singularis de ambiguitatibus*, transmitted in D. 34,5,14,6, the testator should have explicitly said so, had such an unusual understanding been his true intention (see [33, p. 229]). If none of the two slaves dies, the debtor can ask for restitution of one of them via the *condictio* as described in the preceding passage D. 12,6,31. Julian's solution clearly assumes *"aut"* to stand for the exclusive disjunction.

3.2.1.3   Legal Consequence

Julian decides that the heir may not claim anything back: *nihil repetet*. Strictly speaking, this phrase only represents his opinion expressed in the *responsum*. The actual legal consequence would only follow from the applicable procedural formula (see [16, p. 240]):

> *Si paret NN AA < certam rem > dare oportere, quanti ea res est, tantum pecuniam iudex NN AA condemnato, si non paret, absolvito.*

The conditions for this *condictio certae rei* are not explicitly given. The whole point about the actions of the *"si certum petetur"*-type lies in the fact that their formulae did not have to state the actual cause alleged by the plaintiff. Hence, it is the conditional *"si paret"* which abstractly represents them and which the Roman lawyer had to individualize based on the facts.[29] With this understanding Julian's brief statement *"nihil repetet"* simply summarizes that the analysis of the facts have led him to deny the required conditions to be present. Should the debtor decide to go to court against Julian's advice, his adversary would probably be relieved of any liability: *absolutus est*. Julian's analysis partially rested on the interpretation undisputed by him of the word *"aut"* as the exclusive disjunction $aut =)-($. There is admittedly some room for debate in this explanation. Julian's conclusion does not follow directly and logically from the analysis. In fact, one could argue that the debtor has not correctly exercised his right to individualize the legacy by delivering one of the two alternatively bequeathed slaves. Death of one or both of them would be equivalent to a de facto choice by fate. Both interpretations are nevertheless logically consistent with the workings of the logical instrument

---

[29]The precise formulation of the *condictio* is not known. The formula quoted here is the version reconstructed by Lenel. See [16, p. 232] and in particular [16, p. 239] for additional comments on the sources and their interpretation. These uncertainties should not affect this paper's main line of arguments, however.

of exclusive disjunction at the heart of the testamentary clause. My main point is that Julian based his reasoning on the understanding of $aut = \rangle-\langle$. The final phrase "*id remanebit in soluto quod superest*" may be seen as the legal explanation or *ratio decidendi*. If so, his explanation is not very detailed. Unlike his approach in the *liber singularis de ambiguitatibus* he does not offer any more insights into his reasoning. He does not refer to any procedural formula either, as I proposed to do above. The reader has to add all these elements himself in order to complete the treatment of the case.

The example shows a pattern common to Justinian's Digest, where cases are frequently treated only sparsely, without reference to the applicable law and frequently without discussing any gaps or ambiguity in the facts presented. Berman described the resulting "narrowness" or "woodenness" in Roman casuistry as what its jurists desired (see [4, p. 139]). Apparently, they expected their readers to be able to add any missing elements by having recourse to their own technical knowledge. Consequently, they could focus on appraising those elements of a case a good lawyer would need to argue before the court. The second example shall illustrate in more detail how this could have worked in practice.

3.2.2   Quintilian

3.2.2.1   Declamation 264

Ne liceat mulieri nisi dimidiam partem bonorum dare. Quidam duas mulieres dimidiis partibus instituit heredes. Testamentum cognati arguunt.	No more than half of an estate may be bequeathed to a woman. A man named two women heirs to half shares. Relatives contest the will.[30]

This is the introduction to the 264th declamation from the *declamationes minores*, a collection of speeches ascribed to Quintilian to educate young Romans in the art of rhetoric.[31] Some of them at least were designed as arguments at imaginary trials. The declamation reproduced is entitled "*Fraus legis Voconiae*" after a *lex* promulgated in 169 BC. The real Voconian Law prohibited women from being instituted heirs to testators from the highest census class. The wording given in the text does not exactly match the available sources on the *lex*.[32] Obviously, the teacher

---

[30]Translation reproduced from [27, p. 169].

[31]Some scholars question Quintilian's authorship. The collection might just be the notes by an unknown person of Quintilian's courses in rhetoric. See the introductory notes in [27, p. 1].

[32]See Gai. 2,274: *Item mulier, quae ab eo, qui centum milia aeris census est, per legem Voconiam heres institui non potest, tamen fideicommisso relictam sibi hereditatem capere potest.*

and his pupil focused on the art of persuasion rather than on the intricacies of the law, frequently using purely fictitious laws or fictitious versions of actual laws as starting points for their exercises.[33]

The speaker sets out with the telling statement that the wording of the law is clear in itself: *Antequam ius excutio et vim legis, quae per se satis manifesta est, intueor [..]*. He then proceeds to denigrate the position of the opposing party who seems to appeal to family feelings and bonds of blood. Apparently, he says, the testator had his reasons to pass over his relatives and give his estate to the two defendants. The plaintiffs have no sufficient valid arguments to upset the will (for which they would have had to select the *querela inofficiosi testamenti*). They can only attack the law cited by "legal quibbles" (*legem calumniantur*). The speaker then rhetorically asks where the law's true intention lies.

3.2.2.2 Formalization

Contrary to Julian's example discussed above, the starting point here is a question of law rather than of facts. There is no keyword in the text that makes an explicit reference to a logical instrument. As it will become apparent later on while discussing the speaker's chain of arguments, his discourse nevertheless hinges on a skilful use of two versions of the disjunction as presented by Proculus. To begin with, the opposing parties detect two different meanings for the cited law. These meanings are best understood by formal equations. Let n denote the number of heirs instituted by the testator where $m \leq n$ signifies the number of female heirs. Let further x_i denote the portion allocated to heir i. Then the law requires either

- $x_1 + .. + x_m \leq \frac{1}{2}$ or
- $x_i \leq \frac{1}{2}$ for all $i \leq m$

to hold true. Declamation 264 is reduced to the simple case $n = m = 2$. Considering at first only the border cases with $x_i = \frac{1}{2}$ leads to the possible outcomes represented in Table 3.2.

Table 3.2 Inclusive and exclusive disjunction in Quintilian

A	B	A)–(B	A ∨ B	Comments
1	1	0	1	Both women receive ½ of the estate each
1	0	1	1	Woman 1 receives ½ of the estate
0	1	1	1	Woman 2 receives ½ of the estate
0	0	0	0	Neither one of them receives anything

---

[33]On this characteristic of the collection, see [7, p. 84; 131].

The variables A and B in the first two columns denote the two propositions "Lady 1 (2) receives ½ of the estate", respectively. The next two columns then represent two variants of the disjunction discussed by the Stoics. An entry "1" in one of these columns signifies that the outcome is "right" in the light of the applicable law. The first row represents the situation at the trial's outset where the two defendants share the estate among themselves. The relatives receive nothing at all which leads them to take legal action. In the case of the second and third row only one lady receives one half of the estate with the remainder being distributed to the relatives. The fourth row represents the situation where both heirs instituted lose their rights to the estate. This outcome is not considered any further in the text as the *querela* was discarded from the start. In short summary, the formalization takes up the concept of the binary code in systems theory, deciding between "right" and "wrong" of human actions.[34] The next section demonstrates how closely Quintilian's speaker knits his arguments around the two alternatives from the second and third row. These arguments eventually lead to his advocating the solution represented by the first row and the fourth column.

3.2.2.3 Chain of Arguments

Having repeated the wording of the Voconian Law, the speaker proceeds to applying the facts at hand to the *lex*:

Quaero igitur ab istis utram eligant, cum qua milint consistere. Neque enim litigant de bonorum parte, set totum arguunt testamentum.	I ask them therefore which of the two they choose, whom they prefer as their adversary. For they are not litigating about part of the estate, they are challenging the entire will.
Incipiamus igitur ab ea quae prior scripta est. Quid in hac parte testamenti vitiosum est? Vetatur plus quam dimidiam partem bonorum relinquere: dimidiam partem patrimoni accipit. Executiemus postea quale sit illud quod consecutum est: interim hoc prius firmum est, nec everti sequentibus potest.	Let us begin then with the lady who is put down first. What is wrong in this part of the will? It is forbidden to leave more than half an estate: well, she gets half the patrimony. We shall examine later what came next; meanwhile this prior item is solid and cannot be overturned by what follows.[35]

The first sentence takes up a procedural argument. The plaintiffs have to file suit against both heirs.[36] The woman whose name appears first in the will might receive one half of the estate (see the second row in the table above). Then both meanings

---

[34]This representation sidelines the question whether normative texts can serve as truth-bearers in propositional logic at all or whether a form of deontic logic would have to be used.

[35]Translation reproduced from [27, p. 171].

[36]Using the *hereditatis petitio* as single action with regards to the entire estate.

of the law are obviously complied with: *hoc prius firmum est*. But why should the second woman no longer receive anything?

Nec video rationem cur id quod illi [the first] *capere licuit huic* [the second] *non liceat, cum in eodem scripta sit testamento.*	[..] and I don't see any reason why what one of them could inherit the other could not, since she is named in the same will.

Obviously, the speaker denies that the order in which the names appear in the will are of any importance: *Manifestum est nihil posse calumniae admittere verba legis ac scriptum.*[37] Therefore, the second and third rows in the table above are interchangeable. Therefore, both heirs have the same right to their share of the estate. As the result, only the first row in the table leads to the "right" outcome.

The speaker then turns to the position of the opposing party: *Nunc peritissimi*[38] *litium homines ad interpretationem nos iuris adducunt*. If their lawyer shares the view that the wordings of the texts, will and law, are sufficiently clear, he cannot but employ some artificial interpretation to further his clients's interests: *Non enim hanc esse legis voluntatem quae verbis ostendatur videri volunt*. The other party's lawyer must claim that the law's true objective lies in the first meaning formalized above: That the group of female heirs may not receive more than half of an estate. After riling his adversary for questioning the intellectual and verbal powers of the legislators, the speaker points out the following:

Et apparet potuisse legum latorem ut si partem demum patrimonii pervenire ad feminas vellet, partem utique viris relinqui, id ipsum cavere; neque id magno aut difficili circuitu effici potuit, sed vel sic scripta lege, ne plus quam dimidia pars patrimonii ad feminas perveniret.	And if the lawmaker had wished that only half an estate go to females and half be left to males in all circumstances, he could obviously have provided just that. No big, difficult roundabout was needed, only a law so framed that no more than half an estate goes to females.[39]

There is no room for looking for an ulterior motive in a law the wording of which is clear and does not refer to this very motive. Still, the speaker does not disregard the legislator's motive altogether. On the contrary, he skilfully employs this argument of his opponent in the favour of his own clients:

---

[37] [27, p. 171]: "The words of the will, the text, cannot admit of any quibble". The stress should lie here on the clarity of the testator's will. The clarity of the law comes next.

[38] *Peritissimi (periti)* are practitioners rather than scholars (see [10, p. 237]).

[39] Translation reproduced from [27, p. 173].

Quid si ne ratio quidem repugnat scripto et verbis legis istius? Quid enim putas voluisse legis latorem cum hoc ius constitueret? Ne feminas nimias opes possiderent, ne potentia earum civitas premeretur. Hoc ergo adversus singulas constituit.	What if logic too is not at odds with the text and wording of the law? For what do you think the lawmaker intended when he laid down this statute? He did not want women to possess too much wealth, for fear the community be weighed down by their power. So he laid down this against individual women.[40]

If there is such a motive, it speaks for the second meaning of the law, limiting the individual shares allotted to female heirs to half of an estate.

Quid si ne ratio quidem repugnat scripto et verbis legis istius? For sure, *ratio* can stand for many different things and Shackleton's translation using the word "logic" may be too suggestive.[41] Nevertheless, the proposed translation underpins the view that the logic of the disjunction *aut* = ∨ from the table's fourth column does fully support the meaning of the law quoted at the text's beginning. Column 4 spans the entire set of "right" outcomes: Either one or both of the defendants receive half of the estate. As the testator names them both equally in his will, both of them share the estate.

3.2.2.4 Legal Consequence

Quintilian's text does not explicitly treat the legal consequence any more than Julian's text did. The speaker simply denies by his arguments that the relatives find any legal ground for their claim in the *lex*. Hence, the conditions for a *hereditatis petitio* are not met and the defendants have to be relieved of any liability: *absolvitas sunt*.

3.2.2.5 Addenda

The declamation includes a few more topics that, even though not entirely relevant to the main line of the arguments, nevertheless offer additional insight into the general theme investigated in this study.

The truth-value table proposed to formalize the case suggested a simplification. In fact, the speaker's quoted passages only used half shares allotted to the heirs. Towards the end of his speech, the speaker addresses the more general situation, where a testator bequeaths to a number of women smaller shares, the sum of which however exceeds one half of his estate. As the speaker favours the second meaning

---

[40]Translation reproduced from [27, p. 175].

[41]See Giaro [10, p. 424] who lists several examples for the use of *ratio iuris* in the Digest before contrasting *ratio* with Ulpian's and others concept of *utilitas*, and [15, p. 793]. In particular, see the latter's explanations in § 221 on the syllogistic argument and its application to interpreting laws.

of the law, he denies any ground for a suit filed under it: *Nam si movetur lis, non hac, ut opinor, lege litigabitur.* The passage shows that the author of the text was fully capable of understanding the implications of the two equations given above and could even conceive different cases on an abstract level.

The speaker adopts a surprisingly modern stance when he declines to look for an ulterior motive beyond a law's clearly stated wording: *Non enim hanc esse legis voluntatem quae verbis ostendatur videri volunt.*

Nam si apud iudicium hoc semper quaeri de legibus oportet, quid in his iustum, quid aecum, quid conveniens sit civitati, supervacuum fuit scribi omnino leges. Et credo fuisse tempora aliquando quae solam et nudam iustitiae haberent aestimationem. Sed quoniam haec ingeniis in diversum trahebatur, nec umquam satis constitui poterat quid oporteret, certa forma ad quam viveremus instituta est. Hanc illi auctores legum verbis complexi sunt; quam si mutare et ad utilitates suas pervertere licet, omnis vis iuris, omnis usus eripitur.	For if in court there is always to be question about laws, what is just in them, what equitable, what convenable to the community, there was no need for laws to be written at all. And I do believe there were times in the past when justice rested on judgment, alone and unsupported. But men's minds pulled it this way and that, and the right course could never be adequately determined; for that reason a fixed pattern was put in place by which we were to live. Those authors of our laws embraced this pattern in words; if this may be changed and perverted to suit particular interests, there goes the whole meaning and use of law.[42]

This passage seems to make a vivid case in favour of positive laws. Despite Noerr's simile of *leges* as "islands of stability" within a sea of unsystematic law quoted earlier, positivism was certainly not the universally accepted reality of Roman jurisprudence.

A final comment may be added on how we should interpret the Declamations as a general source. A number of authors have altogether discarded them as a reliable source for Roman legal discourse. Wenger [32, p. 253] was more circumspect when he asked for a separate test in each case, recognizing that they were used for training attorneys. Stagl [28, p. 1] showed that Declamation 360 does not simply make rhetorical appeals to equity (as many others certainly do) but includes a number of highly technical legal arguments that by their style and thoroughness could have been taken from the Digest itself. Declamation 264 is certainly lighter on the civil law side. The two emotional appeals against the plaintiffs, who might have been rightly passed over by their parent in his will, and against their attorney, who believes himself intellectually above the lawmakers, show their closeness to traditional rhetoric. But large parts of Declamation 264 read like an essential guide to the interpretation of laws. This trait sets it apart from the more purely rhetoric texts in the collection. One can easily imagine the speaker taking a deeper interest in law and having gone on to practice it in his later life. Declamation 264, like others,

---

[42]Translation reproduced from [27, p. 173].

offers the modern reader a view on those details of the Roman legal profession's workings that the Digest commonly excludes as unnecessary (see [28, p. 8]). In the light of this paper's theme, it nicely complements the rigid bareness in Julian's text.

3.3 Conclusions

3.3.1 Comparing the Two Texts

Both texts only briefly introduce the main facts of their case. Iul. Pal. 161 examines a not further specified obligation. Its alternative options have to be carefully identified in the logically correct way to which the keyword "*aut*" in the very text explicitly points. The reader of Declamation 264 must extrapolate the facts from the short introduction and the arguments used in the main body. The disjunction as logical instrument is only implied by the structure of a case but can be recognized through the speaker's line of arguments. The declamation explicitly states the (fictitious) legal basis for the case, though. Iul. Pal. 161 does not include an explicit legal basis but this can readily be derived from the context with the title "*si certum petetur*".

Julian also only hints at the probable legal consequence should his client decide to file a suit against his adversary. In his opinion based on his logical analysis of the facts, the conditions for a successful *condictio* are simply not met. The modern reader has to add all missing elements of the complete legal "program". Some of the missing elements are illustrated by the court proceedings in Quintilian's text. The plaintiffs allege a violation of the Voconian Law. The speaker, i. e. presumably the student in rhetoric, argues for dismissing the suit: *si non paret, absolvito*. He employs purely objective arguments and rejects a purpose-oriented interpretation of the *lex* as neither the wording of the *lex* nor of the will in question present logical contradictions: *si ne ratio quidem repugnat scripto et verbis legis istius*. He does not accept any moral arguments in favour of the plaintiffs, either, as Julian did in D. 34,5,13,6 in order to alleviate the rigidity of logic.

3.3.2 Logic and Roman Casuistry

Berman [4, p. 139] relates that the Roman jurists wholeheartedly rejected Cicero's proposal for a revised *ars iuris*, securely founded on Greek science and remained "suspicious of the applicability of the higher ranges of Greek philosophy" to the practical needs of jurisprudence. This seems largely to represent the general view expressed in the literature on the subject. My own research leads me to agree insofar as the Roman jurists did certainly not use a purely mechanistic approach derived from logic. As far as Julian's writings are concerned, there are hardly any traces of Aristotelian or Stoic formal inference schemes to be detected. On a lower, less conceptual level, logical instruments can nevertheless be integrated with the traditional picture of Roman casuistry on two accounts. Firstly, logical instruments can themselves be part of the case to be analyzed, as the example of the disjunctive statement in Iul. Pal. 161 shows. Secondly, logic may be brought into play by individual jurists when solving specific cases, as the example taken from Quintilian's Declamations shows. It is fair to say that this latter example is by no means an isolated one.[43] There are numerous more examples from Julian's digest falling into this category and which have to be analyzed in more detail before making more precise qualitative and quantitative statements. Following Miquel, I am inclined to agree that Roman jurists – and namely Julian – were able to think "axiomatically" (although I prefer the term "syllogistically" to avoid misconceptions).[44] It is possible but not necessary that they did use logic in their practice (next to, and artfully combined with the more customary rhetorical devices). It even seems to be probable in Julian's case. Judging from the transmitted sources and, with the notable exception of his *liber singularis de ambiguitatibus,* he did not indulge too freely in making use of his technical logical expertise.

Acknowledgments This research was supported by a Forschungskredit of the University of Zurich, grant no. FK-13-011. I presented the results at a workshop held in Zurich on June 28, 2013 dedicated to "Argumentationsstrukturen im talmudischen und römischen Recht".

References

1. M. Armgardt, Salvius Iulianus als Meister der stoischen Logik – zur Deutung von Iulian D. 34,5,13(14), 2–3, in *Liber amicorum Christoph Krampe zum 70. Geburtstag*, ed. by M. Armgardt, F. Klinck, I. Reichard (Duncker & Humboldt, Berlin, 2013), pp. 29–36
2. C. Atherton, *The Stoics on Ambiguity* (Cambridge University Press, Cambridge, 1995)
3. U. Babusiaux, Celsus und Julian zum Edikt si certum petetur, Bemerkungen zu Prozess und "Aktionendenken", in *Dogmengeschichte und historische Individualität der römischen Juristen*, ed. by C. Baldus, M. Miglietta, G. Santucci, E. Stolfi (Alcione, Trento, 2012), pp. 367–431
4. H.J. Berman, *Law and Revolution, The Formation of Western Legal Tradition* (Belknap, Cambridge, MA, 1983)
5. S. Bobzien, Stoic syllogistic. Oxf. Stud. Anc. Philos. **14**, 133–192 (1996)
6. I.M. Bochenski, *Ancient Formal Logic* (North-Holland, Amsterdam, 1951)
7. S.F. Bonner, *Roman Declamation in the Late Republic and Early Empire* (University Press of Liverpool, Liverpool, 1949)
8. E. Bund, *Untersuchungen zur Methode Julians* (Böhlau, Köln, 1965)

---

[43] Within Quintilian's collection, Declamations 250 and 263 also contain a logical background. Three out of 145 preserved declamations are admittedly a small number.

[44] In my view, the term "axiomatic thinking" evokes the example of Euclidian geometry, a comparison that is not well-suited to Roman jurisprudence but refers to later developments during the seventeenth and eighteenth centuries (see for instance Christian Wolff).

9. L. De Ligt, A philologist reads the Digest: D. 34,5,13(14), 2–3. Tijdschrift vor Rechtsgeschiedenis **66**(1–2), 53–63 (1988)
10. T. Giaro, *Römische Rechtswahrheiten, Ein Gedankenexperiment* (Klostermann, Frankfurt am Main, 2007)
11. M. Kaser, *Zur Methode der römischen Rechtsfindung* (Vandenhoeck & Ruprecht, Göttingen, 1962)
12. W. Kneale, M. Kneale, *The Development of Logic* (Clarendon, Oxford, 1984)
13. W. Kunkel, M. Schermaier, *Römische Rechtsgeschichte*, 14th edn. (Böhlau, Köln/Weimar/Wien, 2005)
14. K. Larenz, *Methodenlehre der Rechtswissenschaft*, 2nd edn. (Springer, Berlin, 1969)
15. H. Lausberg, *Handbuch der literarischen Rhetorik*, 4th edn. (Steiner, Stuttgart, 2008)
16. O. Lenel, *Das Edictum Perpetuum, Ein Versuch zu seiner Wiederherstellung*, 3rd edn. (Tauchnitz, Leipzig, 1956)
17. D. Liebs, *Hermogenians iuris epitomae, Zum Stand der römischen Jurisprudenz im Zeitalter Diokletians* (Vandenhoeck & Ruprecht, Göttingen, 1964)
18. N. Luhmann, *Soziale Systeme* (Suhrkamp stw, Frankfurt am Main, 1987)
19. N. Luhmann, *Das Recht der Gesellschaft* (Suhrkamp stw, Frankfurt am Main, 1995)
20. J. Miquel, Stoische Logik und Jurisprudenz. Zeitschrift für Rechtsgeschichte (Romanistische Abteilung) **87**, 85–122 (1970)
21. D. Noerr, Spruchregel und Generalisierung. Zeitschrift für Rechtsgeschichte (Romanistische Abteilung) **89**, 18–93 (1972)
22. D. Noerr, *Rechtskritik in der römischen Antike* (Bay Akademie Wiss, München, 1974)
23. R. Patterson, *Aristotle's Modal Logic, Essence and Entailment in the Organon* (Cambridge University Press, Cambridge, 1995)
24. A. Scalia, B.A. Garner, *Making Your Case, The Art of Persuading Judges* (Thomson/West, St. Paul, 2008)
25. B. Schmidlin, *Die römischen Rechtsregeln, Versuch einer Typologie* (Böhlau, Köln/Wien, 1970)
26. F. Schulz, *Prinzipien des römischen Rechts* (Duncker & Humblot, Leipzig, 1934)
27. D.R. Shackleton Bailey, *Quintilian. The Lesser Declamations*, vol. 1 (Harvard, London, 2006)
28. J.F. Stagl, Durch Rede zum Recht am Beispiel von Quint. decl. 360. J. Eur. Hist. Law **4**(1), 1–8 (2013)
29. P. Stein, *Regulae iuris: from juristic rules to legal maxims* (Edinburgh University Press, Edinburgh, 1966)
30. J. Stroux, *Römische Rechtswissenschaft und Rhetorik* (Stichnote, Potsdam, 1949)
31. W. Waldstein, Konsequenz als Argument klassischer Juristen. Zeitschrift für Rechtsgeschichte **92**, 26–68 (1975)
32. L. Wenger, *Die Quellen des römischen Rechts* (Holzhausen, Wien, 1953)
33. F. Wieacker, Amoenitates Iuventianae, Zur Charakteristik des Juristen Celsus. IURA **13**, 1–21 (1962)
34. M. Winkler, Zu Logik und Struktur in Julians liber singularis de ambiguitatibus. Zeitschrift für Rechtsgeschichte **130**, 203–233 (2013)
35. P. Ziliotto, Si hominem aut fundum non dederis, centum dari spondes. Studia et documenta historiae et iuris **76**, 392–333 (2010)

Part II
Leibniz, Law and Logic

Chapter 4
Presumptions and Conjectures in Leibniz's Legal Theory

Matthias Armgardt

Abstract This paper focuses on the role of presumptions and conjectures in Leibniz's legal theory. Both presumptions and conjectures are closely connected to the question of the burden of proof for presumptions lead to a shift of the burden. Thus, these notions play an essential role in the practice of litigation: the odds to win a given case are stacked against the party that has to bear the burden of proof. The paper analyses three different texts which bear upon the topic and stem from different phases of Leibniz's production: the *Elementa Juris Naturalis* (1671), the *De legum interpretatione, rationibus, applicatione, systemate* (1678/1679), and the *Nouveaux Essais sur l'entendement humain* (1704, published posthumously in 1765). The aim of the paper is to elucidate how Leibniz developed a subtle theory of legal presumptions and conjectures. Moreover, the paper attempts a first formal reconstruction of this theory.

4.1 Introduction

Gottfried Wilhelm Leibniz's Theory of Law is very important for the history of jurisprudence because of two reasons: First, he was one of the greatest logicians and philosophers of all time; second, he was a highly qualified jurist with profound experience in theory and in praxis of the law of his time. This double-qualification has no counterpart in history. In spite of this, the importance of Leibniz as a master of jurisprudence has not adequately been recognized in Legal History and Legal Philosophy until today.[1] This is due to at least three reasons: First, in his time neither logic nor its application to law was very highly estimated; second Leibniz published only very few papers; third, his legal theory (just as his philosophy) is more complex

---

[1]Fortunately, new studies of the jurisprudence of Leibniz and translations of important texts have appeared recently, e.g. [3, 4, 6, 8–10, 17–19, 31–33].

M. Armgardt (✉)
Full Professor of Civil Law, Ancient History of Law, Roman Law and Modern History of Private Law, Universität Konstanz, Konstanz, Germany
e-mail: matthias.armgardt@uni-konstanz.de

© Springer International Publishing Switzerland 2015
M. Armgardt et al. (eds.), *Past and Present Interactions in Legal Reasoning and Logic*, Logic, Argumentation & Reasoning 7, DOI 10.1007/978-3-319-16021-4_4

and formal than most others, and the interpreter has to consider both logic and the Roman Law of the seventeenth century at the same time.

In this paper, I will focus on the roles of presumptions and conjectures in the legal theory of Leibniz. Both notions are closely related to the notion of burden of proof, since they are essential for the practice of litigation. In fact, the odds to win a given case are stacked against the party that has to bear the burden of proof. Thus, the burden of proof is a crucial point in the law of evidence.

We will analyze three texts by Leibniz, which deal with presumptions and conjectures and stem from different periods of his philosophical production:

- Elementa Juris Naturalis (short: EJN), 1671 [13], VI 1, pp. 431 sqq][2]
- De legum interpretatione, rationibus, applicatione, systemate, 1678/1679 [13], VI 4 C, pp. 2782 sqq]
- Nouveaux Essais sur l'entendement humain, 1704 (published post mortem, 1765) [13], VI 6]

4.2   Presumptions in the *Elementa Juris Naturalis*

The earliest text by Leibniz we examine is part of the Elementa Juris Naturalis (EJN) and was written between 1669 and 1671 during Leibniz's stay in Mainz. As the title shows, Leibniz tries to do for law what Euclid did for mathematics.[3] Leibniz did not publish the EJN. They were the philosophical core part of his great plan for a reform of the Corpus Iuris Civilis, the collection of Ancient Roman Law made by the jurists of the Emperor Justinian in the sixth century. This code was applied in Leibniz's time as Ius Commune.

It is well known that the EJN deals with problems of deontic logic.[4] The main idea of Leibniz is to reduce the deontic modalities (*justum, injustum, debitum, indebitum*) to alethic modalities (*possibile, impossibile, necessarium, contingens*) by making use of the concept of *vir bonus* [13, VI 1, pp. 465 sqq].

In the following parts of the EJN, Leibniz analyses the relations between the juridical modalities (Theoremata quibus combinantur Iuris Modalia inter se) [13, VI 1, pp. 468 sqq] and the relations between the juridical and logical modalities (Theoremata quibus combinantur Iuris Modalia Modalibus Logicis seu justum com possibili) [13, VI 1, pp. 470 sqq]. This has already been carefully examined by Kalinowski and Gardies ([11, pp. 98 sqq]; see also [15, pp. 322 sqq]).

---

[2]For a German translation of parts of the EJN, see [8], pp. 89 sqq].

[3]The roots of this plan probably go back to Leibniz's teacher Erhard Weigel (1625–1699), who was a professor at the University of Jena and wanted to apply the Euclidean method to all branches of knowledge. For more information, see [22, pp. 38 sqq].

[4]For more information, see [29, p. 329], [25, pp. 16 sqq], [11, pp. 79 sqq], [15, pp. 320 sqq] and [16], especially Sect. 4.5.

What has not been analysed from the perspective of legal theory[5] so far is the following part: Theoremata quibus combinentur justum cum existente (Theorems on combining what is just with what is existent) [13, VI 1, pp. 471 sqq].[6] In this part of the EJN, Leibniz frequently uses the verb "praesumere" and the noun "praesumtio"; that is, he deals with legal presumptions.[7]

4.2.1 How to Distinguish Facilius, Praesumitur and Probabilis

The chapter "Theoremata quibus combinentur justum cum existente"[8] starts as follows (13, VI 1, p. 471):

Actus facilius est justus quàm injustus. Item
Actus praesumitur justus.

An action is easier just than unjust. Also
an action is presumed to be just.

Leibniz deals with the qualification of actions as just or unjust. It is presumed that an action is just.[9] It would seem that there is a presumption of justice, because to be just is easier (*facilius*) than to be unjust. But this is wrong as the end of the chapter demonstrates. To avoid a wrong impression at the very beginning of our analysis we have to start at the end of the chapter (13, VI 1, p. 472): Leibniz deals with the three main notions: *facilis* (easy), *probabilis* (probable) and *praesumendi* (it is to be presumed):

Restat discrimina facilis, probabilis, praesumendi explicemus.

Now we have to explain the differences between "easy," "probable" and "it is to be presumed".

Leibniz starts by providing a clarification of the term *facilius* (easier):

Facilius est quod est per se intelligibilius, seu quod pauciora requirit.

Easier is what per se is more intelligible, that is what requires less.

---

[5]Recently, Blank [5] wrote about presumptions in the EJN from a philosophical point of view, stressing the difference between ontological and logical requirements. About presumptions in general, see [7, p. 426] and [1, p. 202]. For a first introduction to the juridical role of presumptions referring to DLI, see [9, pp. xxxiv–xxxvi].

[6]Leibniz later deleted the heading. For a German translation, see [8, pp. 264 sqq].

[7]Busche [8, pp. 265 sqq] translates "praesumitur" as "im Voraus annehmen". Praesumere is a well defined legal terminus technicus and means "vermuten" in German.

[8]Leibniz later deleted this headline.

[9]Burckhardt [7, pp. 425 sqq] explains this theorem by making use of the idea of privation. See also [5, p. 214].

Facilius is a comparative term in relation to the logical form. Its logical form is simpler than the logical form of the opposite. Leibniz goes into detail about this aspect in the further course of the text.

What is the difference between *facilius* and *probabile*? Leibniz explains:

> Probabile est, quod est absolute intelligibilius seu, quod idem est, possibilius. Unde ad probabilitatem requiritur non tantùm facilitas existendi, sed et facilitas coexistendi caeteris impraesentiarum.

> Probable is what is absolutely intelligible or (which is the same) more possible. Because of that, probability requires not only the facility to exist, but also the facility to coexist with other circumstances.

The difference between *facilius* and *probabilis* is that *probabilis* requires more: not only to be more intelligible than the contrary, but also the possibility of coexistence with all other things. The idea of coexistence seems to be a forerunner of the famous concept of compossibility (of individuals) that we find in the texts of the mature Leibniz.[10]

Leibniz continues:

> Ideò generatim definiri nihil potuit de probabilitate, constat enim probabilitas ex collectio omnium circumstantiarum: non potest ergo indefinitè asseri actum probabilius videri justum quàm injustum.

> Thus it is not possible to define something about the probability in general (generatim); the probability consists namely of the collection of all circumstances: thus it is not possible to assert without further determination (indefinitè) that an action is more probable just than unjust.

If we compare this with the very beginning of the chapter (*Actus facilius est justus quàm injustus. Item Actus praesumitur Justus*) we can see the great difference between *facilis* and *probabilis*: there is no contradiction between the beginning and the end of the chapter, because *facilis* and *probabilis* are not the same! The difference lies in the coexistence that we have to take into account if we speak about probability.

There is a second pitfall concerning the interpretation of the beginning of the chapter (13, VI 1, p. 471):

> Actus facilius est justus quàm injustus. Item
> Actus praesumitur Justus.

"Item" does not mean "thus" or "consequently".[11] This is also evident at the end of the chapter (13, VI 1, p. 472):

---

[10]For more information about Leibniz's concept of compossibility, see [28, p. 962] and especially Wilson's [34, p. 1]. Her new ideas considers [24, pp. 105–6]. For a logical analysis of Leibniz's compossibility (especially GP 3, 573), see Lenzen ([15, pp. 327–8] and [14, pp. 185 sqq]). Lenzen draws our attention especially to the connection of the concepts of "compossibility" and "possible world" and thinks that compossibility is not transitive. The transitivity of Leibniz's compossibility is defended by [21, 27, p. 77] and [23, p. 55].

[11][8] translates "item" as "Das bedeutet:". This is also misleading.

Facilius autem et praesumendum differunt ut Minus et pars. Facilius enim est in quo minora vel pauciora quàm in opposito requiruntur, praesumendum cuius requisita requisitorum oppositi pars sunt.

But "easier" and "to be presumed" differ as minus and part. "Easier" namely is what requires less than or minus than the opposite; "to be presumed" is whose requisites are a part of the requisites of the opposite.

After these definitions, Leibniz concludes:

Omne ergo praesumendum est facilius, non contra. Quia etiam omnis pars est minor toto, non omne minus est pars majoris. Sed de his exquisitius alio loco.

Thus everything to be presumed is easier, but not vice versa. Also because every part is smaller than the whole, (but) not every smaller thing is part of a bigger thing. But more about this elsewhere.

This means that at the beginning of the chapter "item" must not be translated as "consequently" or "thus". It is the other way round:

An action is presumed (*praesumitur*) to be rather just than unjust and because of this it is easier (*facilius*) just than unjust.

If Pr(Ja) means "action a is presumed to be just" and Fac(Ja > Ia) "action a is easier just than unjust",[12] then (4.1) holds:

$$Pr\,(Ja) \rightarrow Fac\,(Ja > Ia) \tag{4.1}$$

But (4.2) is, according to Leibniz, not true:

$$Fac\,(Ja > Ia) \nrightarrow Pr\,(Ja) \tag{4.2}$$

At first glance, this seems to be astonishing. Leibniz gives no explanation. A possible reason could be that legal presumptions are only made if there is a special need for them in a legal context, e.g. in case there is a problem to prove something in specific situations.

---

[12]Thus, "Pr" is a kind of propositional operator that produces propositions out of propositions. Leibniz did not develop an explicit semantics for them, nor is this developed in the present paper. Leibniz deals with this notion on the one hand syntactically and on the other hand he deploys some intuititive semantics. Similar applies for "Fac". In relation to "J" and "I", one could treat "Fac" as an operator as well, even though it looks more natural to formulate it as a kind of predicate defined on actions. One way to formulate such kind of predicates is to make use of a more expressive language containing an infinite number of type of objects, such as in Martin-Löf's Type Theory (see [20]). In such a setting we could formulate the following *formation-rule* for "J": J(x): proposition (x: Action) – in words: J(x) becomes a proposition provided x is substituted by a token of the type *action* – or alternatively by *an element of the set of actions*. For the general purposes of the present paper, the full development of such an underlying type-theoretical approach is not necessary. However, the reader might assume some formal typing system underlying the use of J while qualifying actions.

4.2.2 Presumptions That Actions Are Just (Justum) and Not Obligatory (Indebitum)

At the beginning of the chapter Leibniz proves two theorems (13, VI 1, p. 471):

> Th. 1 Actus facilius est justus quàm injustus.
> Item Actus praesumitur justus.

> An action is easier just than unjust.
> Also an action is presumed to be just.

and

> Th. 2 Actus est facilius indebitus quàm debitus.
> Imò actus praesumitur indebitus.

> An action is easier not obligatory than obligatory.
> Even more[13] an action is presumed to be not obligatory.

From a philosophical point of view, these theorems are very important. If an action is presumed to be just, it is allowed, if it is presumed to be not obligatory, it can be omitted. Thus, both theorems contain a strong defence of freedom. We will come back to this point in the next chapter.

4.2.2.1 Presumption That Actions Are Just

Let us now study the arguments of Leibniz. His proof of the first theorem

> Actus facilius est justus quàm injustus. Item Actus praesumitur justus.

runs as follows. The first part deals with the *facilius*-relation:

> Quia facilius evenit aliquid possibile quàm impossibile esse. Nam ad possibile nihil requiritur quàm ut supponatur; ad impossibile verò ut dum supponatur, eius simul oppositum supponatur. Plura ergo requiruntur ad impossibile quàm ad possibile. Ergo facilius est actum esse justum quàm injustum.

> For it is easier for something to turn out to be possible than impossible. For nothing is required for the possible but that it be supposed; for the impossible, however, (it is required) that while it is supposed, its opposite be supposed at the same time. Therefore more things are required for the impossible than for the possible.[14] Hence an action is easier just than unjust.

Here again we find the idea of a simpler logical structure. Leibniz has already made use of it to explain the term *facilius* (cf. Sect. 4.2.2.1). Why is it true that something is easier possible than impossible? It is true, because for the possibility of A you only have to suppose A, but for the impossibility of A you need to suppose

---

[13]"Imò" as item again does not mean "thus".

[14]Translation according to [1, p. 204].

a contradiction consisting of two parts: p and non-p. That is, for the impossible, you need more requirements than for the possible.

To prove that an action is rather just than unjust, Leibniz changes the modalities and analyses the relation of "possible" and "impossible".[15] This shift of modalities is justified by the following theorems of the previous chapters [13, VI 1, pp. 465, 470]:

Iustum/licitum est, quicquid possibile est fieri à viro bono.

Just/allowed is what is possible to make for a good man.

and

Omne justum possibile est.

Every just (action) is possible (for a good man).

By applying these theorems, Leibniz can reduce the deontic modalities to alethic ones. He only has to deal with the question whether the possible is easier than the impossible.

Leibniz continues:

Imò requisita seu supposita possibilis in impossibilis suppositis continentur, non contra.

The requisites or things to be assumed of possible things are contained in the requisites of impossible things, but not vice versa.

In the second part of the proof, Leibniz deals with presumptions.

Praesumitur autem cuius supposita etiam oppositi supposita sunt, non contra. Praesumi igitur est quodammodo praesupponi opposito suo, natura prius esse.

But it is presumed whose requisites are also the requisites of the opposite, not vice versa. For this reason to be presumed is kind of being pre-supposed to its opposite, to be according to nature prior.

This explanation of presumptions shows when presumptions are justified: only when the requisites of the presumed thing are necessarily requisites of the opposite, as well.

Now the proof of the first theorem is perfect:

Ergo Actus praesumitur justus.

Thus an action is presumed to be just.

Let us consider the structure of Leibniz's proof:

---

[15]Blank [5, p. 215] thinks that the "presumption of justice" is a special case of the "presumption of possibility". He refers to [1, pp. 206 sqq]. Leibniz does not make use of the concept "presumption of possibility" in the EJN. He only makes use of the concepts "easier" and "more possible" to justify the presumption that an action is just.

Thesis:
An action a is presumed to be (Pr) just (J):

$$Pr\,(Ja) \tag{4.3}$$

Hypothesis:
An action a is easier (Fac) just (J) than unjust $(\neg J)$[16]:

$$Fac\;(Ja > \neg Ja) \tag{4.4}$$

Just is an action *a* that is possible for a good man (P*):

$$Ja \leftrightarrow P^*a \tag{4.5}$$

Proof of the Hypothesis:
By the combination of (4.4) and (4.5) we obtain:

$$Fac\;(P^*a > \neg P^*a) \tag{4.6}$$

In fact (4.6) follows by substituting P*a for Ja. However, Leibniz provides a less syntactic proof for (4.6) based on the idea that generally for the possible logically less is required than for the impossible, because for the first you only have to defend one statement whereas for the second a contradiction consisting of two parts $(r \wedge \neg r)$ has to be demonstrated.

The problem is that at this stage of the proof we cannot got back to the thesis (4.3). Furthermore, this is not even possible, because according to Leibniz, we cannot conclude from *facilius* to *praesumitur*.[17] This means that in order to complete the proof, we need, in addition to the application of the *facilius*-rule, a juridical justification. However, it cannot be found in this text.

4.2.2.2 Another Proof of the First Theorem

There are manuscript variants regarding the proof of the theorem that an action is presumed to be just.[18] They are worth to be considered[19]:

> Hinc sequitur actum in dubio praesumi licitum. *Praesumitur* enim quicquid est (a) facilius (b) minus. Jam quod facilius < - > est minus opposito suo praesupponendum est, non contra. (a) Idem et probabile est.

---

[16]Recall the afore remark on the operators Pr and Fac.

[17]As already mentioned: "Omne ergo praesumendum est facilius, non contra".

[18][5, p. 216] gives an example.

[19]For the Latin text, see [13, A VI 2, pp. 567 ll. 23 sqq].

Thus in case of doubt, an action is presumed to be permitted. What is easier and less is *presumed*. Thus what is easier and less than its opposite is presumed, not vice versa. And it is probable, too.

Again, Leibniz states that we cannot argue from easiness to presumption.

(b) *Probabile* est quod saepius fit. (a) Jam quidquid est facilius fit saepius (b) Discrimen inter praesumtionem et probabilitatem est quod inter demonstrationem et inductionem (a) scientiam et experientiam. (b) In praesumtione enim ex natura rei demonstramus esse faciliorem ac proinde praesumendam frequentiorem. In probabilitate autem experientiae ope. (a) Quod enim praesumimus, id ex natura sua demonstramus esse facilius, ac proinde praesumimus esse frequentius. Contra probabile inductione scimus esse frequentius atque inde praesumimus esse facilius.

Probable is what happens frequently. And what is easier happens frequently. The difference between presumption and probability is the same as between demonstration and induction, science and experience. To justify a presumption, we show that it is easier by the nature of things and thus presumed to be more frequent. But in probability, we show this with the aid of experience. What we presume is, what we proof to be easier according to its nature, and for this reason we presume it to be more frequent. By contrast, we know by induction that the probable is more frequent and for this reason, we presume that it is easier.

This passage is interesting, because it shows the difference between probability and presumption. The difference can be compared with the difference between science and experience, deduction and induction. To justify a presumption we have to analyse the nature of a thing, whereas probability is based on experience.

Praesumere est (a) incertum pro certo habere in agendo (b) pro certo habere donec oppositum probetur. *Probare* est certum facere, *certum*, cuius veritas clara est. *Pro certo est* quod in agendo sequimur, quasi certum. Praesumendum est quicquid prudenter praesumitur. (. . .) Praesumitur quicquid opposito suo praesupponendum est, non contra.

To presume is to take uncertain things as certain when we act until the opposite is proven. *To prove* is to make sure, *sure* is, whose truth is clear. *To take as certain* is what we take for granted when we act as if it would be certain. What the wise man presumes is to be presumed. (. . .) It is presumed what is to be assumed against its opposite, not vice versa.

Here Leibniz gives a short definition of the term presumption. He makes use of it in the *Nouveaux Essais* much later.

4.2.2.3 Presumption That Actions Are Not Obligatory

Let us now analyse the proof of the second theorem (13, VI 1, p. 471):

Actus est facilius indebitus quàm debitus.
Imò actus praesumitur indebitus.

An action is easier not obligatory than obligatory.
And even more,[20] an action is presumed to be not obligatory.

---

[20]Imò as item again does not mean: thus.

The proof runs as follows:

Quia omne indebitum est justum.[21] Omne debitum est injustum omitti, th. –, jam justum facilius est injusto, imò praesumitur. Ergo indebitum facilius debito; imò praesumitur.

Because every action being not obligatory is just to omit, every action being obligatory is unjust to omit, th. –, every action is easier just than unjust, and it is even presumed. Thus not obligatory is easier than obligatory; and it is even presumed.

Leibniz makes use of theorems of the chapter "Theoremata quibus combinantur Iuris Modalia inter se" [13, VI 1, pp. 468 sqq]. There we find the theorems "Omne indebitum justè omittitur" [13, VI 1, p. 468 line 31] and "Omne debitum est injustum omitti" [13, 1, p. 469 line 1]. By this replacement, Leibniz can reduce theorem 2 of this chapter to theorem 1.

By using modern formalism, Lenzen [15, p. 322] made a logical reconstruction of[22]:

Omne indebitum justè omittitur - et omne quod juste omittitur est indebitum.

as:

$$\neg O\,(\alpha) \leftrightarrow J\,(\neg\alpha) \tag{4.7}$$

And for this reason holds:
Omne debitum est injustum omitti.

$$O\,(\alpha) \rightarrow \neg J\,(\neg\alpha) \tag{4.8}$$

Now Leibniz proves the first part of theorem 2 (13, VI 1, p. 471):

Actus facilius est indebitus quàm debitus.
Indebitum enim est quod justè omittitur, debitum quod injustè, vid. sup th. – Iam justum est facilius injusto; th. – praeced.

An action is easier not obligatory than obligatory.
Because to be not obligatory is what we omit rightfully (justè), obligatory is what we omit not rightfully (injustè), according to a previous theorem. – As has already been proven, to be just is easier than to be unjust, according to the previous theorem.

Again, Leibniz makes use of the already proven substitution of indebitus/debitus by just/unjust. The deontic system works.

Let us now consider how the proofs of theorem 1 and 2 are connected:

Thesis:
An action a is presumed (Pr) to be not obligatory ($\neg O$):

---

[21]The full stop seems to be an error. "Justum" refers to "omitti".

[22]$O(\alpha)$ means "α is obligatory", $P(\alpha)$ means "α is permitted" (Lenzen uses E "erlaubt" instead of P "permitted") and $V(\alpha)$ means "α is prohibited".

$$\text{Pr } (\neg Oa) \tag{4.9}$$

Hypothesis:
An action a is easier (Fac) not obligatory ($\neg O$) than obligatory (O):

$$\text{Fac } (\neg Oa > Oa) \tag{4.10}$$

Proof of the Hypothesis:
Every non-obligatory action is just, every obligatory action is unjust (to omit)[23] and vice versa[24].

$$\neg Oa \leftrightarrow J\neg a \tag{4.11a}$$

$$Oa \leftrightarrow \neg J\neg a \tag{4.11b}$$

By the combination of (4.10) and (4.11) we get:

$$\text{Fac } (J\neg a > \neg J\neg a) \tag{4.12}$$

Now we can refer to the proof of theorem 1:
   Just to omit an action means it is possible, for a good man, to omit the action:

$$J\neg a \leftrightarrow P^*\neg a \tag{4.13}$$

By the combination of (4.12) and (4.13) we get:

$$\text{Fac } \left(P^*\neg a > \neg P^*\neg a\right) \tag{4.14}$$

Afterwards, Leibniz again goes back to his idea that possible is easier than impossible, because of the logical form.

   Two remarks are necessary: First, it is again only the Hypothesis that is proven, not the Thesis, because it is not possible to conclude from *facilius* to *praesumitur*. A juridical justification is once again necessary in addition to it. Second, Leibniz needs the same rules for P*a and for P*¬a, that is for the possibility of an action and for omitting an action. Otherwise, the reference to the proof of theorem 1 would not work.

---

[23] The negative formula $\neg a$ expresses the omission of an action.

[24] We add a negation to both parts of the biconditional: $\neg\neg O(\alpha) \leftrightarrow \neg E(\neg\alpha)$. According to classical logic $\neg\neg O(\alpha)$ is equivalent to $O(\alpha)$.

4.2.2.4   The Philosophical Meaning of Formalism: Defence of Freedom

Leibniz now discusses the philosophical meaning of these theorems (13, VI 1, p. 471):

> Hinc apparet praesumtionem esse pro libertate, pro licentia, pro indifferentia. Contra servitutem, obligationem, determinationem.

> It now appears that there is a presumption in favour of liberty, of permission, of indifference. (And a presumption) against slavery, obligation, limitation.

Thus, the deontic logic is not only a game, but implies and justifies moral values. We have to keep in mind that Leibniz was a jurist at the time of absolutism. These presumptions are not very exciting for jurists of the western world in the twenty-first century. Still, we find here a remarkable philosophical basis for human rights in an area of absolutism.

Leibniz adds:

> Praesumtio est pro minore, pro negante, pro possibilitate, pro duratione; contra maius, contra id quod facti est, contra difficultatem, contra mutationem.

> The presumption is in favour of the minor, of the negation, of the possibility, of continuity; (the presumption is) against the major, the facts, the difficulty, change.

4.2.3   Game-Theory in the EJN

After this, Leibniz makes some remarks against probabilism (13, VI 1, p. 471):

> Sed haec rectè capienda sunt ne cum probabilis quibusdam in abusum torqueantur. Neque enim statim faciliora, probabiliora, praesumenda, etiam sequenda sunt, id est in agendo pro certis habenda à prudente.

> But these considerations have to be understood in the right way, so that they are not perverted in a probabilistic way. Because we should not follow the easier, more probable and presumed (things) straightaway, that is they are not to be taken for sure in the acting of a wise man.

Now Leibniz adds a game-theoretical analysis to recommend reasonable behaviour:

> Ecce enim potest aliquid esse probabilissimum, et tamen si succedat parum fructuosum, si frustretur valde damnosum. Hoc certè nemo prudens suscipiet.

> Behold! Something can be very probable, and in spite of this, if it happens, it is not very fruitful, if it does not happen, it is very harmful. Surely, no wise man will approve of this.

After that Leibniz analyses the opposite situation:

> Contra potest aliquid esse si succedat valde fructuosum, si irritum sit parum damnosum; hîc certè nulla in audacia temeritas erit.

> And vice versa, something can be very fruitful if it happens, but not very harmful if it does not happen. In this case, surely there is no daredevilry to be feared.

Now Leibniz presents the main rule:

Tum demum ergo probabilia sequenda sunt, cum major est ratio probabilitatum quàm effectuum reciprocè, seu si plus probabilior est actus A quàm B quàm melior est effectus B quàm A. Seu si factus ex ductu probabilitatis in bonitatem major est ab A, quàm B.

Thus, the more probable is to be followed only if the benefit of the more probable (event) is bigger than the advantage of the opposite; or if action A is more probable than B, if the benefit of B is less than (the advantage) of A. In other words: if the product of the probability and the benefit is bigger in the case of A than in the case of B.

Now Leibniz gives an example to make sure that the reader really understands the main rule:

Fac ab A probabilitatem esse ut 5, bonitatem ut 4. Factus erit 20. À B probabilitatem esse ut 6, bonitatem ut 3, factus erit 18. Erit ergo A sequendum potius quàm B, etsi minus probabile.

Let A have the probability 5 and the benefit 4. The product is 20. Let B have the probability 6 and the benefit 3. The product is 18. In this case, we have to follow A rather than B, in spite of A being less probable.

Leibniz ends with a moral admonition:

Hinc minimum peccandi periculum maximo etiam commodo proposito vitabit vir bonus, imò et sapiens (nam ut suo loco demonstrabitur, omnis sapiens est vir bonus, quanquam non solus), neque enim maius malum ei evenire potest, quàm ut vir bonus esse desinat.

Because of this, a good man will avoid even minimal danger of sin even if he could gain a very big benefit, all the more a wise man (because it will be shown elsewhere that every wise man is a good man, but not only he alone); because no bigger evil can happen to him than that he ceases being a good man.

4.2.4 Summary

Leibniz applies the concept of presumptions in his deontic logic of the EJN to show that actions are rather just than unjust and rather not obligatory than obligatory. In this context, he distinguishes the concepts of *facilis*, *probabilis* and *praesumitur* and their relations.

4.3 Presumptions and Conjectures in *De legum interpretatione*

Leibniz wrote *De legum interpretatione, rationibus, applicatione, systemate* (short: DLI) probably between 1678 and 1679 in Hannover, that is about 7 years after the draft of the EJN.[25] The thinking of Leibniz had developed brilliantly in the fields of mathematics and philosophy during his stay in Paris (1672–1676) and his visits to London [2, pp. 139 sqq]. After his arrival in Hannover in December 1676 Leibniz came back to law and jurisprudence again [2, pp. 259 sqq]. In DLI Leibniz develops a very rich and practical theory of law worth to be examined very carefully. In this paper I will focus only on the passages about presumptions and conjectures [13, VI 4 C, pp. 2789–10].

Leibniz deals with the *argumentatio probabilis* and makes the following distinction:

> ARGUMENTATIO PROBABILIS procedit vel *a rei naturae*, vel *ab hominum opinione*. A REI NATURAE rursus est vel *Praesumtio* vel *conjectura*. [13, A VI 4 C, p. 2789]

> Probable argumentation comes from either the nature of things or from people's opinions. The former is, in turn, either presumption or conjecture. [9, p. 86]

4.3.1 Presumptions

The first topic concerns presumptions:

> PRAESUMTIO est, si ex his quae vera esse constat necessario sequitur enuntiatio proposita, nullis aliis praeterea requisitis, nisi negativis, ut scilicet nullum extiterit impedimentum. Itaque semper pro eo pronuntiandum est, qui praesumtionem habet, nisi ab altero contrarium probetur. Et tales sunt pleraque rationcinationes in moralibus. (A VI 4 C: 2789)

> It is a *presumption* if the proposed statement [necessarily[26]] follows from what is [. . .[27]] true, without any requirement other than [negative ones[28]], namely that no impediment [for its truth[29]] obtains. Therefore, we will always have to declare ourselves in favour of he who has the presumption unless the other party[30] demonstrates the contrary. Such is most of the moral reasoning.[31] [9, pp. 86 sqq]

Leibniz makes use of a special form of strict implication (*necessario sequitur*). He who has the advantage of the presumption only has to prove the truth of circumstances that strictly imply the controversial fact at issue (*enuntiatio proposita*)

---

[25]I have to mention that I got aware of this excellent text of Leibniz thanks to Marcelo Dascal. We studied his translation of DLI (see [9, pp. 79 sqq]) together in Konstanz, especially the passages about presumptions and conjectures. My analysis of this text is influenced by his ideas and his translation. For the complete Latin text of DLI, see [13, A VI 4 C, pp. 2782–2791].

[26]Dascal's translation omits *necessario* [9, p. 86]

[27]Dascal adds surely.

[28]Dascal: "the negative one" [9, p. 87].

[29]Addition of Dascal.

[30]Dascal: "someone else".

[31]Dascal: "are . . . reasonings".

if there is no impediment (*impedimentum*).[32] But the other party at court has the possibility to prove the contrary and carries the burden of proof.

Let "C" represent the circumstances being true, "F" the controversial fact, "I" the impediment and "□" the symbol of necessity. Then we can write the following, making use of some kind of logical language (such as Constructive Type Theory) where *true* can be expressed at the object language level:

$$\Box\,([C \wedge \neg I] \leftrightarrow F)\ \text{true}\ (\text{provided C}\ true) \tag{4.15}$$

(where *C true* amounts to establish that there is some kind of evidence *a* in Court for C).

We need the bi-conditional, because the party fighting against the presumption wins if it can prove I. We also have to consider that C is certainly true (*vera esse constat*) and it is only uncertain if I or ¬I. The latter means that the party fighting against the presumption has to prove I to win the case – otherwise it loses.

For this reason, concerning the winning strategy, the truth value of F only depends on I or its negation. In other words, since we already know that C is true the debate will only concern the truth or the falsity of I. Thus, from the viewpoint of the strategy the truth value of F depends only on the value of ¬I (provided, as mentioned above that we know C to be true)

$$\neg I \leftrightarrow F \tag{4.16}$$

Dialogical logic might provide a useful tool to develop further this issue. However, such a development is beyond the scope of this paper.

Another possibility to reconstruct the Leibnizean idea of presumptions could be to make use of the non-monotonic logic of presumptions developed by [26, pp. 176–85].

Making use of their inference system IS the impediment "I" would be interpreted with the aid of a meta-rule r_2 as follows [26, pp. 176–8]:

$$r_1 : C \Rightarrow F \tag{4.17}$$

This means that you have to prove C for F. The symbol ⇒ shows that the rule is defeasible (not strict).

$$r_2 : I \Rightarrow \neg\text{appl}\ (r_1) \tag{4.18}$$

This means: if there is an impediment I, r_1 is not applicable or r_1 is overruled by r_2. Further research may show which approach is more fruitful.

A simpler definition of presumptions can be found in the earlier "Definitionum Juris Specimen" (1676):

---

[32]Of course, it is not at all easy to define *impedimentum*.

Praesumtio est, quod pro vero habetur donec contrarium probetur.

Presumption means that something is assumed to be true until the contrary is proven.

Scholz [30, p. 94] points out that Leibniz only has the *praesumtio iuris* or *praesumtio conditionalis* in mind. The *praesumtio absoluta* or *prasumtio iuris* et *de iure* is according to Leibniz no presumption, but a fiction: "Praesumption Juris et de iure est fictio".[33]

4.3.2 Conjectures

Now we have to analyse Leibniz's considerations about conjectures:

> CONJECTURA locum habet, si ad utrumque oppositorum exacte probandum aliqua positiva adhuc requiruntur, quae an vera sint, non constat, pronuntiatur tamen interim pro eo, quod facilius, sive quod pauciora aut in eodem genere minora habet requisite. Atque ita locum habet quod anjunt Jurisconsulti, *semper in obscuris quod minimum est, sequimur.* Huc pertinent doctrina de gradibus probabilitatis, quam nemo quod sciam satis pro dignitate tractavit. (A VI 4 C: 2789–90)

> There is *conjecture* when, in order to prove with accuracy one of two contrary [positions] one has to use some positive [proposition] about whose truth there is no certainty, and nevertheless we meanwhile declare ourselves in favour of what is easiest [to happen], i.e. of that which, in its genus, involves less requisites or smaller ones. This is what the jurists's saying, in the obscure [cases] the minimum must be followed, means. The doctrine of the degrees of probability which, as far as I know, nobody has treated as it should be treated, belongs in here. [9, p. 87]

Presumptions are concerned with uncertain negative requirements; conjectures with uncertain positive ones ("aliqua positiva ... requiruntur, quae an vera sint, non constat").

This means that if it is uncertain whether F or ¬F, we have to apply the *facilius*-rule[34]: We have to assume which is easier to happen, i.e. what has less requisites or smaller ones – unless the other party proves the contrary.

Let $R_{1-n}(F)$ be the requisites for F, $R_{1-m}(\neg F)$ be the requisites for ¬F, n < m and Conj (A) mean "A is assumed meanwhile", then[35]:

$$\text{Conj (F) iff } n < m \tag{4.19}$$

(where n is the number of requisites for F and m for ¬F)

---

[33]We find the same in a Latin letter of Leibniz to Werlhof, Prof. of Law in Helmstedt, written on 17 July 1696 [13, A I 12, pp. 740 sqq] translated in [9, p. 350]: "Every presumption of what is false (which is ordinarily called a legal and de jure [presumption] and usually believed not to admit contrary proof) is a fiction. I do not admit fictions in natural law".

[34]According to Leibniz, this rule can be found already in the Corpus Iuris Civilis (D. 50,17,1,9): "Semper in obscuris quod minimum est, sequimur".

[35]We only consider fewer requisites, not smaller ones.

4.3.3   Comparison of DLI and EJN

Now we have to compare the concepts of presumption and conjecture in DLI with those of EJN.

First of all, we have to consider that the concept of presumption in the EJN is applied to the level of law, whereas DLI only deals with uncertain facts at issue in a lawsuit: in EJN, it is considered if an action is just/unjust or obligatory/not obligatory; in DLI, if something has happened or not.

As we have already seen, the notion presumption in EJN is linked with the *facilius*-rule, but does not result from it (Actually it is the other way round). In DLI, Leibniz does not link his presumption-rule with the concept of *facilius* at all. According to DLI, Leibniz considers only uncertain negative circumstances (*impedimentum*).

But he makes use of the *facilius*-rule to define the concept of conjecture by dealing with uncertain positive requirements: in case of conjectures, it is assumed what is easier (*facilius*) and the notion of *facilius* is very similar (perhaps identical) to the notion of *facilius* in EJN.

Leibniz develops his earlier model of EJN in the DLI: he defines presumptions and makes use of the *facilius*-rule to define conjectures. The relation between presumptions and conjectures is clarified by making use of the concepts of (uncertain) positive and negative circumstances.

4.4   Presumptions and Conjectures in the *Nouveaux Essais*

In the Nouveaux Essais sur l'entendement humain, written in 1704 and directed against Locke once again, the old Leibniz comes back to presumptions and conjectures (cfr. [30, pp. 94–5]):

> Quant à la presomtion, qui est un terme des Jurisconsultes, le bon usage chez eux le distingue de la conjecture. C'est quelque chose de plus, et qui doit passer pour verité provisionellement, jusqu'à ce qu'il y ait preuve du contraire, au lieu qu'un indice, une conjecture doit estre pesée souvent contre une autre conjecture. [...] Presumer n'est donc pas dans ce sens prendre avant la preuve, ce qui n'est point permis, mais prendre par avance mais avec fondement, en attendant une preuve contraire. (*Nouveaux Essais* IV, chap. XIV, 4)

In this context, Leibniz does not present such a highly sophisticated definition as in the DLI. He repeats the easy definition of presumptions he already mentioned in a variant of EJN (1671) [13, A VI 2, p. 567] and in the "Definitionum Juris Specimen" (1676). This is not surprising at all. He only reacts to the lazy use of language by Locke in a non-juridical context. Thus, a deeper analysis is not necessary. The Latin definitions in DLI seem to be the climax of Leibniz's contributions to presumptions and conjectures in law.

4.5 Summary

Leibniz developed a subtle theory of presumptions and conjectures step by step.

In EJN, he makes only use of the concept of presumption and distinguishes it from the concepts of *facilius* and *probabilis*. The *facilius*-relation does not imply legal presumptions, but without the *facilius*-relation there is no presumption. According to Leibniz it is presumed that actions are just and not obligatory.

In DLI, Leibniz defines presumptions and conjectures and analyses their relation by using the concepts of positive and negative circumstances. The definition of presumptions is different from the definition in EJN. To define conjectures, Leibniz makes use of the concept of *facilius* again.

In the *Nouveaux Essais*, Leibniz defines presumptions and conjectures, but does not go as deeply into the subject matter as he has done in ENJ and DLI. Again, we see that the most thorough texts of Leibniz are written in Latin, not in French.

Acknowledgments I want to thank Marcelo Dascal, who draw my attention to this aspect of the legal theory of Leibniz and Shahid Rahman, who made important comments on the draft of this paper.

References

1. R.M. Adams, *Leibniz: Determinist, Idealist, Theist* (Oxford University Press, New York, 1994)
2. R.M. Antognazza, *Leibniz. An Intellectual Biography* (Cambridge University Press, Cambridge, 2009)
3. M. Armgardt, *Das rechtslogische System der Doctrina Conditionum von G. W. Leibniz* (Elwert, Marburg, 2001)
4. A. Artosi, B. Pieri, G. Sartor, *Leibniz: Logico-Philosophical Puzzles in the Law* (Springer, Dordrecht, 2013)
5. A. Blank, Leibniz and the presumption of justice. Stud. Leibnitiana **38**, 209–218 (2006)
6. P. Boucher, Leibniz: What Kinds of Legal Rationalism? in *Leibniz: What Kind of Rationalist?* ed. by M. Dascal (Springer, Dordrecht, 2008), pp. 231–249
7. H. Burckhardt, *Logik und Semiotik in der Philosophie von Leibniz* (Philosophia Verlag, Munich, 1980)
8. H. Busche, *Gottfried Wilhelm Leibniz. Frühe Schriften zum Naturrecht* (Meiner, Hamburg, 2003)
9. M. Dascal, *G.W. Leibniz: The Art of Controversies* (Springer, Dordrecht, 2008)
10. C. Johns, The grounds of right and obligation in Leibniz and Hobbes. Rev. Metaphys. **62**, 551–574 (2009)
11. G. Kalinowski, J.L. Gardies, Un logicien déontique avant la letter: Gottfried Wilhelm Leibniz. Archiv für Rechts- und Sozialphilosophie **60**, 79–112 (1974)
12. G.W. Leibniz, in *Die Philosophische Schriften* (Bd. 7), ed. by C.I. Gerhardt (Weidmann, Berlin, 1874–1890)
13. G.W. Leibniz, in *Sämtliche Schriften und Briefe*, ed. by the Deutsche Akademie der Wissenschaften zu Berlin (Berlin, 1950 sqq). Cited by Series (Reihe) and Volume (Band)
14. W. Lenzen, *Das System der Leibniz'schen Logik* (De Gruyter, Berlin, 1990)
15. W. Lenzen, *Calculus Universalis. Studien zur Logik von G. W. Leibniz* (Mentis, Paderborn, 2004)

16. W. Lenzen, Leibniz on Alethic and Deontic Modal Logic, in *Leibniz et les Puissances du Langage*, ed. by D. Berlioz, F. Nef (Vrin, Paris, 2005), pp. 341–362
17. K. Luig, Leibniz's Concept of Jus Naturale and Lex Naturalis – Defined 'with Geometric Certainty', in *Natural Law and Laws of Nature in Early Modern Europe: Jurisprudence, Theology, Moral and Natural Philosophy*, ed. by L. Daston, M. Stolleis (Ashgate, Farnham, 2008), pp. 183–198
18. S. Magnier, *Approche dialogique de la dynamique épistémique et de la condition juridique* (College Publications, London, 2013)
19. S. Magnier, S. Rahman, Leibniz's Notion of Conditional Right and the Dynamics of Public Announcement, in *Limits of Knowledge Society*, ed. by D.G. Simbotin, O. Gherasim (Iasi-University-Publications, Iasi, 2012), pp. 87–103
20. P. Martin-Löf, *Intuitionistic Type Theory*. Notes by Giovanni Sambin of a Series of Lectures Given in Padua, June 1980 (Bibliopolis, Naples, 1984)
21. B. Mates, *The Philosophy of Leibniz* (Oxford University Press, Oxford, 1968)
22. C. Mercer, *Leibniz's Metaphysics: Its Origins and Development* (Cambridge University Press, Cambridge, 2001)
23. F. Mondadori, Leibniz and the doctrine of inter-world-identity. Stud. Leibnitiana **7**, 21–57 (1975)
24. O. Nachtomy, *Possibility, Agency and Individuality in Leibniz's Metaphysics* (Springer, Dordrecht, 2007)
25. H. Poser, *Zur Theorie der Modalbegriffe bei G. W. Leibniz* (Franz Steiner, Wiesbaden, 1969)
26. H. Prakken, G. Sartor, Presumptions and Burdens of Proof, in *Proceedings of the Nineteenth Annual Conference on Legal Knowledge and Information Systems (JURIX)*, ed. by T. Van Engers (IOS, Amsterdam, 2006), pp. 176–185
27. N. Rescher, *The Philosophy of Leibniz* (Englewood Cliffs, New Jersey, 1967)
28. D. Rutherford, J. Messina, Leibniz on compossibility. Philos. Compass **4**(6), 962–977 (2009)
29. H. Schepers, Zum Problem der Kontingenz bei Leibniz. Die beste der möglichen Welten, in *Collegium philosophicum, Studien, Joachim Ritter zum 60. Geburtstag*, ed. E.-W. Böckenförde et al. (Schwabe, Basel, 1965), pp. 326–350
30. O.R. Scholz, Verbindungen zwischen allgemeiner Hermeneutik und Methodenlehre des Rechts im 17./18. Jh, in *Entwicklung der Methodenlehre in Rechtswissenschaft und Philosophie vom 16. bis zum 18. Jahrhundert. Beiträge zu einem interdisziplinären Symposion in Tübingen, 18.-20. April 1996*, ed. by J. Schröder (Steiner, Stuttgart, 1998), pp. 8–100
31. A. Thiercelin, On Two Argumentative Uses of the Notion of Uncertainty in Law in Leibniz's Juridical Dissertation About Conditions, in *Leibniz: What Kind of Rationalist?* ed. by M. Dascal (Springer, Dordrecht, 2008), pp. 267–278
32. A. Thiercelin, Epistemic and Practical Aspects of Conditionals in Leibniz's Legal Theory of Conditions, in *Approaches to Legal Rationality*, ed. by D.M. Gabbay, P. Canivez, S. Rahman, A. Thiercelin (Springer, Dordrecht, 2011), pp. 203–215
33. E. Vargas, Contingent Propositions and Leibniz's Analysis of Juridical Dispositions, in *Leibniz: What Kind of Rationalist?* ed. by M. Dascal (Springer, Dordrecht, 2008), pp. 267–278
34. C. Wilson, Plenitude and compossibility. Leibniz Rev. **10**, 1–20 (2000)

Chapter 5
Suspensive Condition and Dynamic Epistemic Logic: A Leibnizian Survey

Sébastien Magnier

Abstract In line with Armgardt, Thiercelin carefully studies the Leibnizian notion of suspensive condition—notion that Leibniz sometimes names *moral condition*. Thiercelin points out Leibniz' will to provide a rigorous definition of that kind of condition. Leibniz not only establishes a link between the legal notion of condition and the logical notion of condition, but he also grasps the problematic of suspensive condition through its epistemic and dynamic features. In this paper we start from Thiercelin's reflections about Leibniz' suspensive condition. Thiercelin's work offers an inventory of different clauses that a logical conditional must fulfill to capture the legal meaning of suspensive condition. Our aim is to compare such a definition with the Public Announcement operator semantics of *Dynamic Epistemic Logic* taking advantage of both its model theoretic semantics and its dialogical semantics. We show that the public announcement operator entails the same dynamic and epistemic features than the ones that Leibniz requires with its notion of suspensive condition.

5.1 Introduction

In some of his early writings, Leibniz studies the notion of suspensive condition, notion that he also names *moral condition*. Instead of providing an umpteenth definition of legal notion of condition, Leibniz explores this notion from a new side.[1] Leibniz' original idea is to consider suspensive condition as a part of a conditional proposition, i.e., as the antecedent of a logical conditional—we refer to such a particular conditional as a *suspensive conditional*. By grasping the notion of suspensive condition through a suspensive conditional, Leibniz takes advantage

---

[1] See [4], VI i 101 ; 370.

S. Magnier (✉)
University of Lille 3, UMR 8163 Savoirs, Textes, Langage, 3 Rue du Barreau, 59650
Villeneuve-d'Ascq, France
e-mail: sebastien.magnier@bbox.fr

© Springer International Publishing Switzerland 2015

M. Armgardt et al. (eds.), *Past and Present Interactions in Legal Reasoning and Logic*, Logic, Argumentation & Reasoning 7, DOI 10.1007/978-3-319-16021-4_5

of logical tools developed for conditional propositions. Accordingly, the meaning of a suspensive conditional may be expressed, compared and studied in relation to conditional propositions.

Traditionally in law, suspensive conditions were (and in some extent continue to be) considered through the notion of existence. According to Leibniz, this onto-logical problem should be turned into an epistemo-logical problem.[2] Even if some propositions are either true or false, their truth values may be ignored, unknown. It is precisely what happens with suspensive condition: the contracting parties (must) ignore whether the condition (the antecedent of the suspensive conditional) is true or false.[3] Either way, they can learn the truth value of this condition thanks to what Leibniz names a *certification*.[4] After that certification, contracting parties know whether the condition is true or not. Therefore suspensive condition becomes a matter of a dynamic epistemic problem instead of a question about the existence of that condition. However in Leibniz' days, the development of logic was not as rich as today, and that is why he mainly uses ideas from propositional logic (even if his work leads to new paths in direction of modal logic and probabilities).

5.1.1 Suspensive Conditional: Example and (Logical) Definition

For didactic reasons, we base our reflections on one example. Let us take the following proposition stated by Primus in favour of Secundus:

If a ship arrives from Asia, I give you 100 coins.

From this proposition a conditional obligation emerges between Primus and Secundus. According to Leibniz, this conditional obligation can be reconstructed using the conditional proposition of the form[5]:

Proposition 1 If *a ship arrives from Asia* then *Primus has to give 100 coins to Secundus.*

From a logical point of view, "Primus has to give 100 coins to Secundus" represents the consequent (B) of the conditional Proposition 1 and this consequent is dependent on the satisfaction of the antecedent (A), i.e., "A ship arrives from Asia". Now from a legal point of view, Primus means that he will give the coins to

---

[2]Leibniz' contribution takes on two sides: a logical one and an epistemological one. That is why we claim that he turned the ontological problem into an epistemo-logical one.

[3]On one hand if one of the contracting parties knows that the condition is already fulfilled and yet uses it as a suspensive condition, he harms the one who is still ignorant. On the other hand if the condition is known to be fulfilled it is no longer a suspensive condition and thus the obligation can not be considered as conditional anymore.

[4]This Leibnizian concept will be presented and discussed in the next sections.

[5]From now on, we change the typesetting when we use words which refer to a logical constant.

Secundus only if a ship is arrived from Asia. Hence, as long as the truth value of the proposition A is not known (i.e., the proposition stating that a ship is arrived is true or the proposition stating that a ship is arrived is false), the truth value of the proposition B remains suspended. The truth value of the consequent B is implied by the satisfaction of the antecedent A but it also depends on that satisfaction. This leads Leibniz to the definition below—this definition is brought to light in [13]:

Definition 1 (Leibnizian Suspensive Conditional) A suspensive conditional is a conditional such that the antecedent implies and suspends the consequent. *Then we have:* A `implies and suspends` B.

If we consider the suspensive condition as a conditional proposition, the legal link between the antecedent and the consequent is provided by the definition above. Although the notion of implication is common in logic, the notion of *suspension* does not really correspond to any (already existing) formal concept. However, the notion of suspension plays a key role in this conditional relation. Then the problem is the following: how can we logically formalize this notion of suspension and the entailed specific conditional? Leibniz' solution is twofold. First, he gives some logical conditions and then he adds other pragmatic conditions. Thiercelin [13] provides an inventory of different clauses that sum up these conditions. We use these clauses to provide the Definition 2:

Definition 2 (Clauses) A conditional relation satisfying all these clauses corresponds to the Leibnizian definition of a legal conditional.

 (i)   The consequent cannot be true if the antecedent is not true.
 (ii) The consequent cannot be its own condition.
 (iii) The truth value of the consequent cannot be known as long as the truth value of the antecedent is not certified.
 (iv) If it is certain that the antecedent is true then the consequent is also true.
 (v)   If it is certain that the antecedent is false then the consequent is also false.
 (vi) The antecedent cannot be a contradiction.
 (vii) The consequent cannot be a tautology.

Clauses (ii) and (vii) specify the logical nature of the consequent whereas the logical nature of the antecedent is specified by clause (vi). The truth conditions of this specific conditional relation are defined by clauses (i, iii, iv and v).

These clauses will help us to grasp which kind of logical conditionality is the most suitable to capture the Leibnizian notion of suspensive condition. In fact, if several conditionals seem to be appropriate, most of them cannot meet all these clauses. In this paper we do not intend to investigate the whole Leibnizian system, since it is already remarkably made in [1][6] and [13–15]. We propose here to reconsider Leibniz' view about the dynamic and epistemic features of suspensive condition using on one hand some of Thiercelin's reflections and on the other hand some logical tools

---

[6]Armgardt's work launched a host of new researches on Leibniz' approach to legal rationality.

recently developed in the epistemic field. Recent works [6, 9] have shown that some concepts of *Dynamic Epistemic Logic* (**DEL**)[7] are very close to Leibniz' logical intuitions about the dynamic and epistemic features of suspensive conditional. It is precisely this relation between suspensive conditional and **DEL** we explore in this paper. First we focus on the logical conditional relation and consider different kinds of conditional connectives showing how/why they fail to capture the meaning of the Leibnizian suspensive conditional. Then we underline the importance of the epistemic and dynamic features entailed by the suspensive conditional. The last part is devoted to public announcement operator. We show why *Dynamic Epistemic Logic* is relevant in the context of the Leibnizian conception of suspensive condition.

5.2 Some Disappointing Conditionals...

In this section we mainly focus on some logical conditional relations showing how/why they fail to capture the conditional relation entailed by a suspensive conditional—as it is defined through the clauses presented in the previous section.[8]

5.2.1 Classical Conditionals

We start presenting the material conditional and then we discuss the bi-conditional. We show that their respective conditional relation do not correspond to the one required for the suspensive conditional.

5.2.1.1 The Material Conditional

The most naive way to formalize the suspensive conditional is to use the material conditional relation "if A then B" such as the Proposition 1 suggests, but without any other restrictions. Thus we get the Proposition 2 below:

---

[7]**DEL** is an umbrella term which refers to different logical approaches to information change using epistemic events such as announcement or action. Here we use *Public Announcement Logic* (**PAL**). See [17] for an overview of these logics. Note that the term *public* simply means that all agents receive the announcement at the same time and all of them know that they receive the announcement at the same time.

[8]We do not consider strict conditional. As [1, pp. 140–141] pointed out, a legal condition is a disposition, that is to say, the conditional relation between the antecedent and the consequent comes directly from an act of will (from at least one person). Since an act of will is always contingent, a legal condition cannot be correctly formalized using a relation such as: "it is necessary that if p then q".

Proposition 2 If *a ship arrives from Asia* then *Primus has to give 100 coins to Secundus.*
Formally we have: $A \rightarrow B$.

At first sight, this if ... then relation seems to satisfy the conditional relation. Nevertheless, it does not met the requirements given by Leibniz. No thorough study is required to discern the huge gap between the material conditional and the Leibnizian suspensive conditional. It is enough to consider the truth values of a material conditional relation in order to be convinced. The Proposition 2 is true if the proposition A is false whatever the truth value of the proposition B might be. Thus the truth value of the consequent B does not depend on the truth value of the antecedent A. Consequently this conditional proposition entails no particular dependent relation between the antecedent and the consequent. For all that, from the legal point of view, Primus creates (due to the conditional obligation) a relationship of dependence between the action "give the coins to Secundus" and the fact "a ship is arrived from Asia" but this dependent relation cannot be established by a material conditional.

In fact, from a logical point of view, the truth values of a proposition $A \rightarrow B$ are equivalent to the truth conditions of a disjunction such that $\neg A \vee B$. If we transpose this equivalence to the conditional obligation taken by Primus and Secundus, we obtain the Proposition 3:

Proposition 3 Either *it is* false *that a ship arrives from Asia* or *Primus has to give 100 coins to Secundus.*
Formally we have: $\neg A \vee B$.

The Proposition 3 neither corresponds to Primus' will nor to the legal link expressed in Leibniz' definition.

5.2.1.2 The Bi-conditional

We have just seen that material conditional cannot correctly formalize a suspensive conditional because it establishes no link between the antecedent and the consequent of the conditional. A simple solution could be to force a stronger link by doubling the conditional relation. This solution leads to the Proposition 4:

Proposition 4 If *A ship arrives from Asia* then *Primus has to give 100 coins to Secundus* and if *Primus has to give 100 coins to Secundus* then *a ship arrives from Asia.*
Formally we have: $A \leftrightarrow B$.

Nevertheless, since $A \leftrightarrow B$ is equivalent to $(A \rightarrow B) \wedge (B \rightarrow A)$, this solution entails (at least) two problems:

1. It preserves the shortcoming of material conditional,
2. It requires pragmatic conditions.

First, let us assume that the propositions A and B are false. For the same reasons given above, the Proposition 4 is trivially true. Let us now consider that both propositions are true. In this case, even if the first conjunct "If a ship arrives from Asia then Primus has to give 100 coins to Secundus" seems to be reasonable, the second conjunct "If Primus has to give 100 coins to Secundus then a ship arrives from Asia" sounds odd: the fact that Primus has to give 100 coins to Secundus appears as being the condition of the arrival of the ship. But this fact cannot cause the arrival of a ship from Asia. Moreover, it cannot be the condition of the suspensive conditional. Instead, it is its outcome, its conclusion.

The intuition behind the will to use the bi-conditional is probably the following: Primus has to give the coins to Secundus if and only if a ship arrives from Asia or because a ship is arrived from Asia. To do this, we need to add pragmatic conditions (at least) stating that the second conjunct must be evaluated once the first conjunct is satisfied. Now, the semantics of the bi-conditional—using the classical conjunction ∧—does not integrate such a temporal dimension.[9] Moreover, a condition stating that the first conjunct must be satisfied only if the antecedent (i.e., proposition A) is true is required.[10] If the aim of the bi-conditional is to establish a strong relation between the antecedent and the consequent, this requires pragmatic conditions.

5.2.2 More Refined Conditionals

Classical conditionals are not appropriate to formalize Leibniz' suspensive condition. There are other possibilities which consist to use a convertible conditional or a connexive conditional.

5.2.2.1 The Convertible Conditional

The particular relation between the antecedent and the consequent of a suspensive conditional we are looking for seems to require a kind of convertibility. Indeed, when Primus says that he will give 100 coins to Secundus if a ship arrives from Asia, he will not give him the coins if no ship arrives. It is exactly what expresses the convertibility principle displayed in the Table 5.1.

Table 5.1 Convertibility principle

Convertibility principle: $(\varphi \rightarrow \psi) \rightarrow (\neg\varphi \rightarrow \neg\psi)$

[9] A sequential conjunction might be more appropriate for this purpose since it introduces an implicit temporal dimension. Another strategy could consist in improperly adding an index in order to make a temporal distinction between different propositions or use a temporal modal logic.

[10] Otherwise the proposition becomes trivially true.

Table 5.2 Limits of the convertibility

1. $(A \rightarrow B) \rightarrow (\neg A \rightarrow \neg B)$	Convertibility principle
2. $(A \rightarrow B) \rightarrow (B \rightarrow A)$	Contra-position on the consequent 1
3. $(B \rightarrow A) \rightarrow (\neg B \rightarrow \neg A)$	Convertibility on the consequent 2
4. $(B \rightarrow A) \rightarrow (A \rightarrow B)$	Contra-position on the consequent 3
5. $B \rightarrow B$	Transitivity 4

Table 5.3 Second connexive thesis

Second Boecian connexive thesis:	$(\varphi \rightarrow \psi) \rightarrow \neg(\varphi \rightarrow \neg\psi)$
Second Aristotelian connexive thesis:	$(\varphi \rightarrow \psi) \rightarrow \neg(\neg\varphi \rightarrow \neg\psi)$

This principle states that if the antecedent of a convertible conditional is false then the consequent is also false—what seems to perfectly fit Leibniz' analysis. Indeed Leibniz' pragmatic conditions force a kind of negative link. This negative link can be clarified through the Proposition 5:

Proposition 5 If if *a ship arrives from Asia* then *Primus has to give 100 coins to Secundus* then if no *ship arrives from Asia* then *Primus has not to give 100 coins to Secundus*.
Formally we have: $(A \rightarrow B) \rightarrow (\neg A \rightarrow \neg B)$.

This convertibility principle satisfies clauses (iv) and (v) and could be used to tighten the relation between the antecedent and the consequent. The problem is that if the convertibility principle tightens the link between the antecedent and the consequent, that link becomes too strong and leads to the fact of having B as its own condition—see Table 5.2.

In accordance with the convertibility principle and its consequence (Table 5.2), the Proposition 6 follows from the Proposition 5:

Proposition 6 If *Primus has to give 100 coins to Secundus* then *Primus has to give 100 coins to Secundus*.
Formally we have: $B \rightarrow B$.

Obviously, it is not what a suspensive conditional means: the antecedent and the consequent must differ from each other. Otherwise, the conditional obligation would be trivial because it would ever be valid and consequently it would be of no interest from a legal point of view. Thus, although the convertibility principle satisfies clauses (iv) and (v), it infringes the clause (ii). Therefore a convertible conditional cannot help us to correctly formalize a suspensive condition and does not correspond to the suspensive conditional we are looking for.

5.2.2.2   The Connexive Conditional

The connexive thesis, presented in Table 5.3, may be used as a softened version of the convertibility principle—this solution corresponds to Thiercelin's choice.

This approach includes (at least) two advantages.

1. It allows to have a strict link between the antecedent and the consequent of a conditional proposition without making them convertible, so the consequent cannot become its own condition. Thus, clauses (iv) and (v) are fulfilled without infringing the clause (ii).
2. The semantics of the connexive conditional entails specific requirements for its antecedent and its consequent: they must be contingent propositions. Hence, a contradiction cannot be the antecedent of a connexive conditional and a tautology cannot be the consequent of such a conditional. This second point implies the satisfaction of clauses (vi) and (vii).

Consequently, the connexive conditional represents a reasonable choice to logically formalize the Leibnizian suspensive conditional.

However, the underlying logic to the connexive conditional is not classical; it is not even an extension of classical logic. On one hand, the connexive conditional satisfies clauses which are infringed by other conditional connectives, but on the other hand we must give up on classical logic. Furthermore, the connexive conditional says nothing about the epistemic characteristics and totally neglects the dynamic features pointed out by Leibniz, especially the certification act.

As we have shown, the material conditional is the worst logical conditional to model suspensive conditional. In its turn, the bi-conditional leads to undesirable difficulties. The convertible conditional allows to tighten the relation between the antecedent and the consequent but there is a price to pay: the consequent may become its own condition. The connexive thesis offers a good compromise between the material conditional and the convertibility principle. But once more there is a cost: we give up on classical logic. Moreover the connexive conditional expresses nothing concerning the epistemic characteristics of the condition and its dynamic features.[11] In fact all of these conditional connectives miss the point of Leibniz' original analysis of the suspensive condition, namely, its epistemic and dynamic features.

5.3   Suspensive Condition and Epistemic Default

In this section, we show that the epistemic and dynamic features of the suspensive conditions play a key role in the suspensive condition in Leibniz' analysis and even in the (French) contemporary Civil Law. We start by briefly outlining some characteristics shared by the Leibnizian conception of suspensive condition and its legal

---

[11]We can easily fill the epistemic gap by adding an epistemic modality K in the logical language— which is proposed by Thiercelin. But even if this solution fills up the epistemic gap, it still cannot internalize the dynamic aspects of the suspensive condition.

definition in the French Civil Code. Then we focus on the Leibnizian certification and its epistemic value, which we compare to the announcement operator in **PAL**.

5.3.1   The Epistemic Default and the French Civil Code

We underline the relevance of Leibniz' analysis of the suspensive condition in terms of an epistemic default, by showing how his ideas are still valid today. Indeed in the [3], there are some marks which seem to be very close to Leibniz' dynamic epistemic considerations, and the clauses (i–vii) can be literally or substantially found in different articles.[12]

5.3.1.1   The Importance of the Ignorance

It is very interesting to point out the fact that, concerning the suspensive condition, the French Law also stresses on the epistemic default. The Article 1168 defines the notion of *obligation conditionnelle* as being an obligation depending on a future and uncertain event—Table 5.4.

The Article 1181,[13] which defines the legal notion of suspensive condition is consistent with Leibniz' analysis of the epistemic default—Table 5.5. This article states that if the event is a future one, it must be uncertain, i.e., not known yet, and if the event is a past one, it must not be already known. Therefore, the event must be first and foremost "unknown".[14]

The condition (i.e., the antecedent of a suspensive conditional) may indifferently be a past event or a future event, the main important point is that this event is not already known as certain by the contracting parties. They must not know (i.e.,

Table 5.4 Article 1168 French Civil Code

Art. 1168	*An obligation is conditional where it is made to depend upon a future and uncertain event, either by suspending it until the event happens, or by cancelling it, according to whether the event happens or not*

---

[12]The translation of the articles come from http://www.legifrance.gouv.fr/Traductions/Liste-des-traductions-Legifrance.

[13]In [7] we discuss this article.

[14]The Doctrine does not completely adhere to this lecture. Nowadays this article is subjected to discussions because some jurists reject past events as condition (See for example [12, p. 1131]). These discussions have led to rewriting projects divided in two opposite streams: one which rejects past events (and by, the epistemic default is deleted) and another which consecrates the epistemic default. We are currently studying this topic and the results will be the subject of future paper.

Table 5.5 Article 1181 French Civil Code

Art. 1181	*An obligation contracted under a condition precedent is one which depends either on a future and uncertain event, or on an event having presently happened, but still unknown to the parties*
	In the first case, the obligation may be performed only after the event
	In the second case, the obligation takes effect as from the day when it was contracted

Table 5.6 Article 1176 French Civil Code

Art. 1176	*Where an obligation is contracted subject to the condition that an event will happen within a fixed time, that condition is deemed failed where the time has elapsed without the event having happened. Where there is no time fixed, the condition may always be fulfilled; and it is deemed failed only when it has become certain that the event will not happen*

Table 5.7 Article 1177 French Civil Code

Art. 1177	*Where an obligation is contracted subject to the condition that an event will not happen in a fixed time, that condition is fulfilled when that time has expired without the event having happened: it is so too when, before the term, it is certain that the event will not happen; and where there is no determined time, it is fulfilled only when it is certain that the event will not happen*

ignore) whether the condition already is or eventually will be fulfilled or not. This event is uncertain or unknown only relatively to contracting parties' knowledge. Hence, according to Leibniz and the French Law, the suspensive condition takes root in an epistemic default.[15]

5.3.1.2 Clauses (v), (vi) and Articles 1176, 1177, 1172

Some similarities with the clause (v) may be found in Articles 1176 and 1177 (respectively Tables 5.6 and 5.7).

Article 1176—Table 5.6—states that "*an obligation is contracted subject to the condition that an event will happen [...] it is deemed failed only when it has become certain that the event will not happen*". Article 1177—Table 5.7—is about conditional obligation taken under negative event: "*an obligation is subject to the*

---

[15]It is not the case in every contemporary law. For example, in the Quebec Civil Code, the Article 1498 states that "*An obligation is not conditional if it or its extinction depends on an event that, unknown to the parties, had already occurred at the time that the debtor obligated himself conditionally.*"—see [2].

Table 5.8 Article 1172 French Civil Code

Art. 1172	*Any condition relating to an impossible thing, or contrary to public morals, or prohibited by law, is void, and renders void the agreement which depends upon it*

condition that an event will not happen [...] [*the condition*] *is fulfilled only when it is certain that the event will not happen*". In both cases, the obligation does not hold if it is certain that the condition (the antecedent) is unfulfilled (false), what sounds similar to clause (v).

We also find some similarity with the clause (vi) in the Article 1172—Table 5.8. This article states that any convention whose the condition (i.e., the antecedent) is impossible is nil. By definition, a contradiction represents a logical impossibility, hence a condition of this type automatically renders nil the suspensive condition.[16]

The closeness between Leibniz' notion of suspensive condition and its French definition is interesting and deserves to be further investigated. This could provide logical tools for those who aim to formalize legal reasoning dealing with suspensive conditions.

5.3.2 From Certification to Public Announcement

First, we consider the epistemic value of the certification, i.e., why and how the certification may change the knowledge; and then we show that it fits with **PAL** announcement process.

5.3.2.1 The Certification and its Epistemic Value

According to Leibniz, the antecedent of a suspensive conditional must be unknown, e.g., Primus and Secundus must ignore whether a ship is already arrived or will arrive one day. Nevertheless, this does not mean that the proposition "a ship arrive from Asia" is undetermined. This proposition has a truth value: it is either true or false. It is just not possible for Primus and Secundus (when they contract the conditional obligation) to determine whether its truth value is true or false. Thus it is not the truth value of the proposition in itself which defaults: it is the knowledge that the contracting parties have about this truth value. They remain ignorant until a special *event* breaks their epistemic default. It is this special event that Leibniz names *certification*. The certification reveals the truth value of a proposition. Thanks

---

[16]This point was also admitted by Roman jurists who called such a condition a "ridiculous" condition. In fact, it does not make any legal sense to formulate a suspensive conditional involving a logical falsity.

to the certification, the contracting parties' uncertainty concerning the truth value of the condition (i.e., the antecedent of the suspensive conditional) turns to be certain. If before the certification they were ignorant, after the certification they know whether the proposition is true or false. This certification changes contracting parties' knowledge but it also determines the truth value of the consequent of a suspensive conditional. For example, if it is certified that a ship is arrived from Asia then Secundus knows that Primus has to give him the coins; but if it is certified that no ship is arrived from Asia then Secundus knows that he cannot legally claim for the coins. Consequently, this epistemic feature deserves to be explicitly introduced in the logical formulation, thus we reformulate Proposition 1 via Proposition 7.

Proposition 7 If *a ship arrives from Asia* then *Secundus* knows *that Primus has to give him 100 coins.*
Formally we have: $A \rightarrow K_s B$.

Even if the Proposition 7 makes explicit the epistemic features of the suspensive condition (using operator K), the dynamic features of the certification are still lacking. However, the certification cannot be dissociated from the epistemic features of suspensive condition. It is the certification which gives rise to the knowledge change. We claim that on this very point, Leibniz' certification seems to work exactly like an epistemic event as it is conceived in **DEL** and more precisely like a public announcement.

5.3.2.2 Certification, Announcement and Epistemic Dynamics

Contrary to the different conditional relations discussed in Sect. 5.2, the public announcement operator introduces an epistemic dynamics. In **PAL**, if a proposition is truthfully and successfully announced, agents learn the truth value of this proposition, i.e., if before an announcement agents may ignore the truth value of a given proposition, after its announcement they know that the announced proposition is true.[17] Announcements change the agents knowledge: after learning something true, they know it.[18] By making public the truth value of the announced proposition, a public announcement operator modifies the agents' knowledge.[19]

---

[17]In Sect. 5.4.1.2 we further explain what "successfully announced" means.

[18]That means they no longer consider it as being (possibly) false. Due to this learning mechanism, we can consider that an announcement implicitly introduces a temporal dimension: *before* the announcement the agents may ignore what is announced, but *after* the announcement they can no longer ignore it.

[19]This works only for successful propositions, namely propositions where the truth value may not be influenced by its announcement (non-epistemic propositions). However if we only consider factual events—as the [3] seems to suggest—unsuccessful updates will not come up. They may arise only when we announce epistemic propositions. Moreover, factual announcements always

Now according to Leibniz, the certification of the condition can be made if and only if the condition "exists" or is true, exactly like a public announcement.[20] In Primus and Secundus example, the arrival of the ship renders the proposition: "a ship arrives from Asia" true. But as it is underlined in [14], Leibniz pointed out in his Definition 52 that "parfois l'on dit que la condition est en suspens lorsqu'elle a certes existé ou fait défaut, mais que cela est encore incertain, auquel cas il voudrait mieux dire qu'elle est suspendue".[21] The fulfilment of the condition is not enough to be certain that it exists, i.e., to know that the condition is fulfilled. Thiercelin continues saying "il se peut donc qu'une condition existe [le navire est arrivé d'Asie] mais que son événement ne se soit pas encore produit au sens où l'on ne sait pas encore [...] que la condition existe".[22] In other words, it is possible that the condition exists/is fulfilled, but without an event to reveal its existence, we remain ignorant of that.

PAL public announcement operator allows us to accurately reconstruct Thiercelin's quotation in the following terms: it is possible that the proposition of the announcement operator is satisfied—it is true that a ship from Asia is arrived— but since the announcement is not made, this fact (a ship is arrived from Asia) is not commonly known by agents (Primus and Secundus). Indeed an announcement requires two steps:

1. The proposition is true, i.e., there is a fact corresponding to "a ship is arrived from Asia",
2. The truth of this fact is publicly announced.

Otherwise, as long as the proposition "a ship is arrives from Asia" is not publicly announced, Primus and Secundus' knowledge do not change, they remain ignorant of the arrival of the ship. And then according to Primus and Secundus' knowledge (since they ignore whether a ship is arrived or not) they cannot make a difference between:

1. No ship is arrived (the condition is not fulfilled), and
2. A ship is arrived (the condition is fulfilled) but this fact is not publicly announced.

In both cases, the condition remains uncertain according to their knowledge. The fact "a ship is arrived from Asia" must be publicly announced to change their respective knowledge.

---

entail a *common knowledge* of the announced fact. A formula φ is commonly known if and only if "everybody knows that everybody knows that... everybody knows φ" and is written $C\varphi$. Since this point is directly related to the clause (ii), we further discuss it in Sect. 5.4.1.2.

[20] Only truthful announcement can be made. See Sect. 5.4.1.2.

[21] Quotation extracted from Leibniz' Definition 52. See [4], A VI i 71-95 or [15], p. 141.

[22] See [14], p. 141.

5.3.2.3   Certification as an Announcement

The considerations we have developed above lead us to establish a strict parallel between on one hand the Leibnizian concept of certification and on the other hand a public announcement. Both are used to reveal the truth value of a given proposition and in both cases this revelation changes the knowledge of the agents who receive it. If before the certification/announcement, the proposition was uncertain—in the sense that agents were unable to determine whether the proposition is true or false—after this certification/announcement, the proposition is not uncertain anymore. They consider the proposition as certain, that is, they know whether the proposition is true or false. This meets Leibniz' Definition 52 and Thiercelin's remark stating that the existence of the condition is not enough. An event—more precisely an epistemic event—is needed. This epistemic event is named certification according to Leibniz' terminology, and it perfectly fits with the public announcement operator.

Using **PAL** announcement operator, we can directly introduce the dynamics entailed by the certification into the logical formalization. So, slightly changing the intended interpretation of a public announcement operator—without changing its logical meaning—the Proposition 7 can be replaced by the Proposition 8.[23]

Proposition 8 If it is certified *that a ship is arrived from Asia* then after the certification, *Secundus* knows *that Primus has to give him 100 coins*.
Formally we have: $[A] K_s B$.

A public announcement has its own semantics and a proposition can be announced only if this proposition fulfils some conditions. Therefore, if we compare these conditions and the semantics of a public announcement operator with the Leibnizian definition of suspensive conditional (Definition 2), we can measure the relevance of the link that we establish between a public announcement operator and the Leibnizian concept of suspensive condition.

5.4   Public Announcement Operator and Suspensive Conditional

A public announcement operator $[\varphi]\psi$ represents an original alternative to the conditional connectives we have discussed in Sect. 5.2.[24] First **PAL** mainly focuses on the dynamic changes of information and their consequences on agents' knowledge. Moreover the semantic definition of a public announcement operator is conditional.

---

[23]Public announcement interpretation would have given: "If it is publicly announced that a ship is arrived from Asia then after this announcement, Secundus knows that Primus has to give him 100 coins".

[24]Let us say that φ is the announcement and ψ its postcondition.

In the previous section, we have shown that the public announcement operator is very close to the Leibnizian notion of certification because of its dynamic epistemic features. In this section, we now compare the different clauses defining the Leibnizian notion of suspensive conditional (Definition 2) with the public announcement operator. First we focus on its model theoretic semantics and next on its dialogical semantics.

5.4.1   From a Model Point of View

Public Announcement Logic comes from a long tradition initiated in [10] and is fully developed in [17].

5.4.1.1   The Model Theoretic Semantics of Announcement

According to the model theoretic semantics of a public announcement—displayed in the Table 5.9—, the evaluation of the formula ψ (in $[\varphi]\psi$) depends on the announcement φ. The announcement ψ in a model $\mathcal{M}$ generates a submodel $\mathcal{M}|\varphi$. Hence, as long as the formula φ is not announced, ψ cannot be evaluated in the submodel $\mathcal{M}|\varphi$.[25] Hence, the antecedent φ implies the consequent ψ; and at first sight it also seems to suspend this consequent (at least as long as the announcement is not made).[26]

As surprising as it may sound, the public announcement operator semantics offers a conditional relation which seems to perfectly fit most of the clauses given in the Definition 2.

Table 5.9 Model theoretic semantics of public announcement operator

$\mathcal{M}, s \vDash [\varphi]\psi$	iff	$\mathcal{M}, s \vDash \varphi$	implies	$\mathcal{M}|\varphi, s \vDash \psi$

---

[25]In the model theoretic setting, after a public announcement, the model is cut into two parts. On one hand there are the situations which satisfy the announcement (i.e., the situations where the announcement is true before its announcement) and on the other hand there are the situations which do not satisfy the announcement (i.e., the situations where the announcement is false before its announcement). The situations which do not satisfy the announcement are deleted from the model and only the situations where the proposition announced was true before its announcement are kept. Removing some situations from the model, the announcement affects the accessibility relations between the situations: some accessibility relations are removed. This mechanism explains why the knowledge of the agents changes.

[26]In order to check if the antecedent actually suspends the consequent, we need to thoroughly consider whether all the different clauses are satisfied or not.

5.4.1.2   PAL Semantics & the Suspensive Conditional

We divide the clauses listed in the Definition 2 in three thematics:

1. Satisfaction conditions of the conditional relation, clauses (i), (iii), (iv) and (v);
2. Nature of the antecedent, clause (vi);
3. Nature of the consequent, clauses (ii) and (vii).

The Conditional Relation

As we have seen, the postcondition of an announcement cannot be evaluated as long as the announcement is not made. This means that the truth value of the postcondition is entailed by the announcement only when the proposition is certified (publicly announced). Therefore the clause (iii) is fulfilled: the truth value of the postcondition cannot be known until the announcement/certification is made.

Let us now focus on clauses (iv) and (v). Both clauses are about the truth value of the consequent. This truth value directly depends on the truth value of the antecedent. If the antecedent is true then the consequent must be true—clause (iv)—conversely, if the antecedent is false then the consequent must be false—clause (v). The clause (iv) is immediately satisfied by the semantics of a public announcement since only truthful announcements can be made. Even if only the propositions can be announce in **PAL**, the case of a false announcement is tackled in [17, p. 106]. If we assume that a false announcement is made, this announcement creates some trouble. After an announcement, only the situations where the announcement is true before the announcement are kept. But, if the announcement is not true in the situation in which it is made, after this announcement, this situation cannot remain in the model anymore. This situation is removed from the model and thus it is impossible to determine a truth value of the postcondition. Indeed, a valuation function is only defined over the contexts belonging to a model. It is in order to avoid such a disagreement that **PAL** only considers truthful announcements.[27] Whereas the clause (iv) is directly satisfied, the clause (v) is indirectly satisfied. It is satisfied because **PAL** cannot technically deal with false announcements. After a false announcement it would not be possible to determine the truth value of the postcondition.

The clause (i) seems to be a bit more difficult to satisfy. This clause stipulates that the consequent cannot be true if the antecedent is not true. The problem is that nothing forbids to have a model in which the postcondition of a public announcement operator is (contingently) satisfied while the announcement is not. For example in the situation 1 of the model of the Fig. 5.1 the postcondition seems to be satisfied because B holds in the situation s_1. However, the semantics of the public

---

[27]Even if false announcements cannot be made, it remains possible to truthfully announce that a given proposition is false. We should not make a confusion between *truthfully* announce that something is false with *falsely* announce that something is true.

Fig. 5.1 Dependance relation

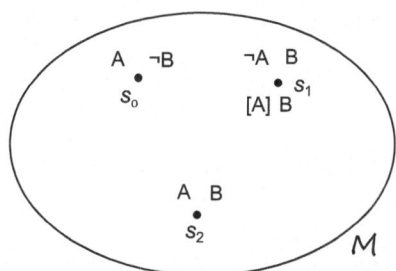

announcement requires that $\mathcal{M}, s_1 \models [A]B$ iff $\mathcal{M}, s_1 \models A$ *implies* $\mathcal{M}|A, s_1 \models B$. Thus the postcondition must be evaluated only in the submodel generated by the announcement. But in the situation s_1 the announcement cannot be made because A is false. In fact in situation s_1 and in accordance with Leibniz' words, we could say that A cannot be certified. Therefore, although B is true, i.e., we have $\mathcal{M}, s_1 \models B$, we cannot have $\mathcal{M}|A, s_1 \models B$.

Hence in any case the postcondition cannot be true if the announcement is not made and the announcement cannot be made if it is not true. Thus the clause (i) is fulfilled: the postcondition cannot be true if the announcement is not true.

Nature of the Antecedent

The clause (vi) forces a condition on the nature of the antecedent. This clause stipulates that the antecedent cannot be a contradiction. If we consider a material conditional it is easy to understand the necessity of this clause. In Sect. 5.2.1.1 we have seen that a material conditional is trivially true if the antecedent is false. By definition (and according to classical logic) a contradiction is always false. Therefore a contradiction as the antecedent of such a conditional would render trivial the conditional relation. Now, in the context of **PAL**, since this logic only allows truthful announcement, it is impossible to announce a contradiction. Consequently, the truthful requirement of **PAL** satisfies the clause (vi): no contradiction can be announced.

Nature of the Consequent

The clause (ii) specifies that the consequent cannot be its own condition. If B is its own condition, (from a legal point of view) the conditional obligation would be meaningless. Moreover it also sounds as a tautology. Nevertheless, $[\varphi]\varphi$ is not a valid formula in **PAL**.

If φ has an epistemic part, the truth of this epistemic part may be influenced by its announcement, that is φ may become false because of its announcement. In such a case the update is said to be *unsuccessful*. In a nutshell, an unsuccessful update

Table 5.10 Necessitation of $[\varphi]$

From ψ infer $[\varphi]\psi$

results from an announcement which is about a fact and an epistemic observation concerning that fact. e.g., "a ship is arrived from Asia and Secundus does not know that a ship is arrived from Asia".[28] Even if it was true that Secundus was ignorant about this fact, as soon as it is certified (publicly announced), Secundus immediately learns that fact. Thus, after the certification, the truthful proposition "a ship is arrived from Asia and Secundus does not know that a ship is arrived from Asia" turns out to be false. Now if we deal only with factual announcements, what seems sufficient in accordance with Article 1168 and 1181 of the [3], we can be certain that such unsuccessful updates will never be produced—because factual announcements do not contain epistemic proposition and always entail common knowledge. In that case the clause (ii) is fulfilled.[29]

Let us now consider the clause (vii) which states that a tautology cannot be the consequent of a suspensive conditional. This clause is more difficult to satisfy with the current treatment of public announcement in **PAL**. In fact, there is an inference rule in **PAL** called "necessitation rule" stating that from a valid formula ψ one can infer $[\varphi]\psi$ which infringes this clause—Table 5.10. From a valid formula ψ, any (true) formula φ can be announced because a truthful announcement cannot change the truth value of a valid formula.

The model theoretic semantics of the public announcement operator is, on this very point, not appropriate and cannot satisfy this clause. Nevertheless the dialogical approach to **PAL** easily solves this problem.

5.4.2   From a Dialogical point of view

The dialogical semantics displayed in the Table 5.11 is provided for the first time in [5] and it is prove to be sound and complete with respect to **PAL** semantics in [8]. Therefore all clauses discussed in the previous section are still satisfied by the dialogical semantics of the public announcement operator. Now the dialogical semantics goes a step further since it can solve the problem mentioned above concerning the clause (vii) and it also offers an argumentative approach to **PAL**.

---

[28]It is obvious that the valuation of an epistemic proposition may change according to the model considered. In any case, such an announcement will ever produce an unsuccessful update. See [16].

[29]Moreover, this leads to the following schemata: "[*factual proposition*] *modal proposition*"— whatever the interpretation of the modality is (epistemic or deontic).

Table 5.11 Particle rule for public announcement operator

Logical constant	X Utterance	Y Challenge ?	X Defense !
$[\varphi]\,\psi$, the defender has the choice	$\mathscr{A}\,\|i : [\varphi]\psi$	$\mathscr{A}\,\|i : ?_{[\,]}$	$\mathscr{A}\,\|i : \neg\varphi$ or $\mathscr{A} \bullet \varphi\|i : \psi$

5.4.2.1 The Dialogical Solution to the Clause (vii)

The dialogical approach to logic is an argumentative approach to logic, that is a two-player game which is used to study and/or investigate different logical systems. This particularity allows more flexibility than the model theory. In fact the dialogical approach to logic requires two kinds of rules: particle and structural rules. Whereas the former provides a local semantics (how players have to/can use each logical constant), the latter furnishes a global semantics (adding condition over the game) and can be seen as a pragmatic semantics.[30] Taking benefits of this pragmatic semantics, the clause (vii) can at this stage be solved. We can add (on the dialogical system used) a structural rule which allows us to modify the normativity of the dialog. Indeed it is easy to formulate a structural rule which forbids to have a tautology as the postcondition of a public announcement operator. Such a rule works as *test* on the postcondition.[31]

For example, let us consider that a player of the dialog claims "$[\varphi]\psi$". After the challenge of the other player, this player can claim that ψ is a satisfiable proposition. According to the structural rule we are discussing, the other player can challenge this claim. Thus the challenger opens a subdialog in which he bears the burden of the proof[32] and tries to verify whether ψ can be falsified or not. If he succeeds, he shows that ψ is not falsifiable, in other words he shows that ψ is a tautology. In that case in accordance with the structural rule, the challenger wins the dialog. Besides, if the challenger fails to show that ψ is a tautology—what means that ψ is a contingent proposition—, he loses the subdialog and the play goes back to the upper dialog. Therefore the subdialog behaves as a filter allowing to reject a tautology as a postcondition; and thus the clause (vii) can be fulfilled.

The clause (vii) was the only one which was not satisfied using **PAL** semantics. If we use the dialogical semantics of **PAL**, all of the clauses from the Definition 2

---

[30] Structural rules allow to formalize conditions which are not purely logical but pragmatic. Nevertheless those conditions modify the underlying logical system (semantics).

[31] This is the solution we adopt in [6, chap.6]. This structural rule is inspired by the F operator, introduced for the first time in 1997 by Rahman in the text of his habilitation (Saarbrücken, 1997), then it has been worked out in [11].

[32] That is, he plays under the formal restriction: he is allowed to utter an atomic proposition only if his adversary has uttered this atomic proposition first.

are satisfied to the extend that the dialogical approach to **PAL** satisfies the clause (vii) and that the satisfaction of the previous clauses is preserved. We now focus on the particle rule for the announcement operator.

5.4.2.2   The Particle Rule for Announcement Operator

The particle rule for the announcement operator provides its local meaning (see Table 5.11), that is, it provides a rule stating how such a constant can be used—i.e., challenged or defended—by players during the dialog. In dialogical logic the two-players framework splits the use of a logical constant in different steps. First, there is a **X**-utterance, next a **Y**-challenge and then there is a **X**-defense.[33]

The particle rule of the public announcement operator works in the same way. First, the player **X** utters $[\varphi]\psi$. Next, it is player **Y**'s turn and he challenges **X**'s utterance by asking whether he rejects or accepts to defend that the announcement is the case. Thus the defender **X** has the choice. He can choose to reject $(\mathscr{A}\,|\,i\,:\,\neg\varphi)$ or accept $(\mathscr{A}\bullet\varphi\,|\,i\,:\,\psi)$ to be committed in such a defense.[34]

The conditional form of the public announcement operator appears in its dialogical semantics as being a matter of choice. Indeed, through its utterance, the player **X** says nothing more than: *"If it is the case that... then..."* what leads his adversary to ask him: *"Is it the case or not?"*. The question of the choice appears clearly in the last step. The defender **X** can answer saying: *"Yes, it is the case that...,* *so..."* or *"No, it is not case that..."*. Let us now see how such a rule works with Primus and Secundus example.

5.4.2.3   From Primus and Secundus' Point of view

In order to provide an intuitive meaning to the particle rule of a public announcement operator, we substitute players **X** and **Y** by Primus and Secundus and we replace φ and ψ by A and $\mathrm{K}_{Secundus}$B. In this case, Primus starts the exchange claiming in front of Secundus:

1. Primus: "If it is certified that a ship arrives from Asia, you know that I have to give you 100 coins."   (utterance)

    $\epsilon\,|\,1\,:\,[\mathrm{A}]\mathrm{K}_{Secundus}\mathsf{B}$

---

[33]Only the particle rule for the negation does not support any defense.

[34]$\mathscr{A}$ represents a list of announced formulas, we write ϵ when the list is empty. So, if the announcement is made, the formula announced is added to this list. A formula added to the list holds for all of the next contextual points of the game. A contextual point i is the dialogical counterpart of a situation in the model theory. A challenge on K-operator is required to move to another contextual point. See [5] and [6, chap. 4–5].

2. Secundus: "Is it certified that a ship is arrived or not?"   (challenge)
   $\epsilon | 1 : ?_{[\,]}$

From now on, Primus has the choice for his answer. He may say:

3.a. Primus: "It is certified that a ship is arrived, so you know that I have to give
     you the coins."   (defense)
     $A | 1 : K_{Secundus} B$
     or
3.b. Primus: "No ship is arrived."   (defense)
     $\epsilon | 1 : \neg A$

If Primus chooses the first option (3.a.), Secundus will assuredly ask him for the
coins but if he chooses the last option (3.b.), saying that no ship is arrived, Secundus
may contradict him, saying that a ship is arrived.[35] Note that if Secundus does that,
he will bear the burden of the proof concerning the arrival of the ship from Asia.
If he succeeds, Primus will be compelled to accept the fact that a ship is arrived.
Thus, in accordance with his initial claim, Primus will have to defend that from
now on Secundus knows that he has to give him the coins. Even if the discussion
between Primus and Secundus starts with the suspensive condition, the exchanges
turn around the ability/will to certify whether the condition (the announcement) is
fulfilled or not. And next, if the condition is fulfilled, it deals with the epistemic
change entailed by this fulfilment.

In the dialogical approach to the public announcement operator, the Leibnizian
certification is interpreted as a player's ability to certify, that is, as the player's
ability to prove that the condition is fulfilled. Conversely the suspension appears
as being an inability to certify, to prove that the condition is fulfilled. This point is
extremely interesting since an argumentative approach to the suspensive conditional
explains its meaning in terms of ability/inability to provide a proof of its condition.
As long as the claimant (Secundus in our example) is not able to prove that the
condition is fulfilled, this player will never (legally) obtain the coins. Besides, if he
is able to provide such a proof, this proof modifies the agents' knowledge during the
game.

5.5 Conclusion

Starting from Thiercelin's analysis of Leibniz' notion of suspensive condition, we
have shown that the dialogical semantics of the public announcement operator
internalizes in its logical language all the different clauses defining the suspensive

---

[35]Thanks to the particle rule for the negation: if **X** utters $\neg\varphi$ then **Y** can only challenge this
utterance with φ and there is no defense for **X**. Player **X** can only counter-challenge φ using the
appropriate particle rule.

conditional (Definition 2). Satisfying all these clauses is not the only advantage of the dialogical approach, it also provides a unique framework allowing the formal combination of:

- The ignorance on the fulfilment of the condition,
- The epistemic dynamics triggered after the fulfilment of the condition,
- The requirement of a public knowledge about the evidence for such a fulfilment.[36]

Moreover this framework offers an interesting dynamic and interactive framework which has (at least) two other advantages:

1. It provides a real understanding of the suspensive condition: the suspensive condition is a fact that a player has to prove during a legal trial; and as long as the proof is not provided this player cannot legally claim for the benefit of the conditional obligation.
2. It brings to light the question of the burden of the proof in the course of a legal trial: if Primus claims that he will give 100 coins to Secundus if a ship arrives from Asia, he can reject the burden of the proof of the arrival of the ship upon Secundus in the case where he asks for the coins.[37]

In fact, the dialogical approach underlines the double dynamics of the suspensive condition: it represents an epistemic event which changes the agents knowledge (epistemic dynamics) and this event has to be proved in the course of a legal trial (argumentative dynamics). Both dynamic dimensions are illustrated in Fig. 5.2.

*
*　*

This Leibnizian survey we have provided shows that the nature of the young Leibniz' interests is not only logical and philosophical. The starting point of his intellectual reflections is grounded in the dynamic nature of the argumentative practice in the process of acquisition of knowledge and decision making.

---

[36]This is probably the most important feature coming from the public announcement operator and it corresponds to the fact that the existence/fulfilment of the condition is not enough (see §Certification, Announcement and Epistemic Dynamics, Sect. 5.3.2). Its existence/fulfilment must be of public knowledge, that is, commonly known—what can only be achieved by a successful public announcement.

[37]This point is interesting and deserves to be further investigated since it fits with the French definition of the burden of the proof in a legal trial: "*A person who claims the performance of an obligation must prove it. Reciprocally, a person who claims to be released must substantiate the payment or the fact which has produced the extinguishment of his obligation.*" [3, Art.1315].

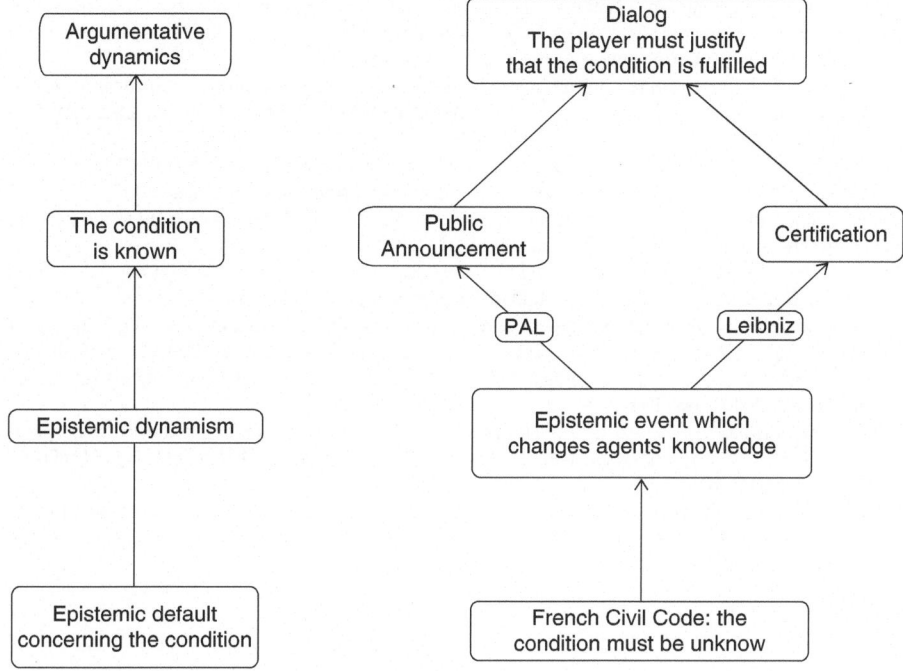

Fig. 5.2 The suspensive condition and its double dynamics

Acknowledgements This paper has been supported by JuriLog Project (ANR11 FRAL 003 01), hosted at the Maison européenne des sciences de l'homme et de la société (MESHS – USR 3185).

References

1. M. Armgardt, *Das rechtslogische System der "Doctrina conditionum" von Gottfried Wilhelm Leibniz* (Elwert, Marburg, 2001)
2. Code Civil of Québec (1991)
3. French Civil Code (1804)
4. G.W. Leibniz, *Sämtliche Schriften und Briefe* (Akademie, Berlin, 1964)
5. S. Magnier, PAC vs. DEMAL, a dialogical reconstruction of public announcement logic with common knowledge, in *Logic of Knowledge. Theory and Applications*, ed. by C. Barès, S. Magnier, F. Salguero (College Publications, London, 2012), pp. 159–179
6. S. Magnier, *Approche dialogique de la dynamique épistémique et de la condition juridique.* (College Publications, London, 2013)
7. S. Magnier, La logique au service du droit: L'analyse de la signification du terme "incertain" dans la d'éfinition de la condition suspensive du droit civil français, Int. J. Semiot. Law (2015). Springer. doi:10.1007/s11196-015-9412-2. http://link.springer.com/article/10.1007/s11196-015-9412-2
8. S. Magnier, T. de Lima, A soundness & completeness proof on dialogs and dynamic epistemic logic, in *Dynamics in Logic*, ed. by P. Allo, F. Poggiolesi, S. Smets. Logique et Analyse (2014) http://www.logiqueetanalyse.be/

9. S. Magnier, S. Rahman, Leibniz's notion of conditional right and the dynamics of public announcement, in *Limits of Knowledge Society*, vol. 2, ed. by D.G. Simbotin, O. Gherasim (2012), pp. 87–103 http://www.euroinst.ro/titlu.php?id=1228
10. J. Plaza, Logics of public communications, in *Proceedings 4th International Symposium on Methodologies for Intelligent Systems*, Charlotte, ed. by M.S. Pfeifer, M. Hadzikadic, Z.W. Ras (1989), pp. 201–216
11. S. Rahman, H. Rückert, Dialogical connexive logic. Synthese **127**(1), 105–139 (2001)
12. F. Terré, Ph. Simler, Y. Lequette, *Droit civil – Les obligations* (Dalloz, Paris, 2002)
13. A. Thiercelin, Conditions, conditionnels, droits conditionnels: L'articulation du jeune Leibniz (première partie). Studia Leibnitiana, **41**(1), 21 (2009)
14. A. Thiercelin, *La théorie juridique leibnizienne des conditions : ce que la logique fait au droit (ce que le droit fait à la logique)*. Ph.D. thesis, Université de Lille, 2009
15. A. Thiercelin, Epistemic and practical aspects of conditionals in Leibniz's legal theory of conditions, in *Approaches to Legal Rationality*, ed. by D. Gabbay, P. Canivez, S. Rahman, A. Thiercelin. Logic, Epistemology, and the Unity of Science (LEUS), vol. 20 (Springer, Dordrecht, 2011), pp. 203–215
16. H. van Ditmarsch, Dynamic epistemic logic, the Moore sentence, and the Fitch paradox (2010). http://arche-wiki.st-and.ac.uk/~ahwiki/pub/Arche/ArcheLogicGroup10Jun2009/slidesMooreFitch.pdf
17. H. van Ditmarsch, W. van der Hoek, B. Kooi, *Dynamic Epistemic Logic*. Synthese Library: Studies in Epistemology, Logic, Methodology, and Philosophy of Science, vol. 337 (Springer, Dordrecht, 2008)

Chapter 6
The Rhetor's Dilemma: Leibniz's Approach to an Ancient Case

Bettine Jankowski

Abstract A well-known puzzle (supposedly) dating back to ancient Greece is Protagoras's court case against his pupil Euathlus. If and how this case can be solved has been under discussion since antiquity. This essay focuses on the approach taken by Gottfried Wilhelm Leibniz and compares it to both traditional and contemporary solutions. It will be argued that Leibniz's way of dealing with the dilemma is more juridical than most solutions, using principles of law to find a way out of the seemingly paradox situation. Other authors in the last hundred years also attempted – very different – solutions for the puzzle, as it will be shown.

6.1 Introduction

Some court cases have been used throughout the ages as examples for both philosophical and legal problems. These 'schoolroom' examples show very well the strong connection both disciplines share: depending on the writer's perspective, the same problem can be seen either from a philosophical or from a legal perspective, leading sometimes to very different solutions. In other instances the same conclusion is reached but using very different methodologies. The best person to consider such examples is someone who has a background in both fields. Therefore, if Gottfried Wilhelm Leibniz writes on a legal case that has a long philosophical history, as he does for example in *De casibus perplexis,* cap. XVI., case VIII [13, VI 1, pp. 241–242], it is a contribution that should be taken into close consideration.

Leibniz held a habilitation in Philosophy and a doctorate in Law; he also worked in both fields and can be considered an authority of his time on matters that concern the connection of law and logic. The following text approaches the 'Rhetor's Dilemma' (as Protagoras's court case against his pupil Euathlus is sometimes called) mainly from Leibniz's perspective and compares it to modern day solutions of this ancient case, while giving the history and sources due consideration.

B. Jankowski (✉)
Fachbereich Rechtswissenschaft, Universität Konstanz, Konstanz, Germany
e-mail: bettine.jankowski@uni-konstanz.de

© Springer International Publishing Switzerland 2015   95
M. Armgardt et al. (eds.), *Past and Present Interactions in Legal Reasoning and Logic,* Logic, Argumentation & Reasoning 7, DOI 10.1007/978-3-319-16021-4_6

6.2   The Case

The probably most famous rendition of the case is given in Aulus Gellius's *Attic Nights,*[1] which is also the text Leibniz most closely relates to in his works on perplex cases.

In *De Casibus perplexis*, Leibniz himself gives a much shorter version of the events than Gellius, which seems to be so specific as to imply that the reader has previous knowledge of the story. For an introduction, Gellius account is more appropriate:

A master and his pupil entered into an agreement according to which the pupil would pay the Master for being taught rhetoric but the fee would only be due after the pupil won his first case in court. The rhetor taught his student according to the contract. However, after the lessons were finished, the student didn't practice his art in court. Since the student didn't take any court cases, the rhetor decided to sue him for the agreed on fee. In court, the rhetor argued his case by stating that he was owed the money no matter the outcome of the case, since either the court decided in his favour, therefore entitling him to the money, or in favour of his pupil, which would fulfil the condition set in the contract, once again making the money due. The pupil disagreed with his teacher's statement and argued that he would not have to pay the fee either way. If he won the court case, by the judges's ruling, he did not owe his teacher the money; if he lost, then the contract's condition would not be fulfilled, freeing him from the obligation to pay the fee.

The question posed can be put as follows: Who should win the case? Or, more specifically: What should the judges decide? Both the rhetor and his pupil refer to similar sounding arguments that seem to be mirror images of each other, leading to much confusion on first reading them.

6.3   Sources of the Dilemma

Before considering possible solutions, it is important to understand the sources the puzzle is traced back to. The setting in all sources is ancient Greece. However, there is no evidence readily available to prove whether the story actually took place. From the Latin sources available, it seems that considering the problem as an actual court case in Athens would be mere conjecture.

---

[1] The original text can be found in Gellius 5.10.9 [16, pp. 200–201]. For an English translation, see [20, pp. 404–409].

6.3.1   Ancient Sources

As mentioned above, the probably best known source for the story is Aulus Gellius's *Noctes Atticae* [16, pp. 200–201]. In this compilation, Gellius refers to the case as an example for reciprocal arguments and calls it "well known". It appears that at the time of Gellius's writing, in the second century AD, this case was already a much used example in rhetoric teaching and philosophy. Gellius attributes the role of the rhetor to Protagoras with his pupil Euathlus playing the counterpart of the student. From its setting, the original court case would have taken place about 600 years before Gellius writings, if it is attributed to the Sophist Protagoras, who lived in the fifth century BC. The judges Gellius mentions in his account, the Areopagites, were Athens's High Court. Had the case actually taken place during Protagoras's lifetime, the law system of ancient Athens would have been relevant to the proceedings since Protagoras is reported to have lived in Athens. However, as there is no source or evidence for an actual court case between the two protagonists and not much is known about procedural law in ancient Greece, a 'law' solution using classical Greek law does not seem to be very promising.

The same story with minor details told differently is given in Apuleius's *Florida* fragment 18 [9, pp. 168–170], which is dated around the same time as Gellius's writing. Both writers knew each other's work and most likely were influenced by each other [10, p. 23].

There is no clear evidence that Protagoras's pupil Euathlus existed much less that the anecdote ever took place. The only available source older than the ones given before is a very short line in Quintilian's *Institutionis Oratoriae Liber III* [1, pp. 27–28], which is dated to the first century BC. It provides roughly the same anecdote, even though it lacks almost all identifying details.

What all these sources have in common is that the case is cited as an example of 'sophism', whereby the term is clearly meant to have a negative connotation.[2]

A different but also well-known version of the story is attributed to Hermogenes. However, the original text where he is supposed to have mentioned the anecdote is lost; the references given even by Leibniz are to the *Prolegomena*, i.e. medieval introductions to forensic rhetoric based on Hermogenes's work.[3] The only difference with respect to the other sources is that here the participants of the court case are not Protagoras and Euathlus but Corax and his pupil Tisias, the alleged founders of rhetorics (cf. [23, p. 1]) – which suits the context of what is taught in the *Prolegomena*.

---

[2] See as an example the ending in Apuleius's telling of the case.

[3] Leibniz cites Johan Sturm's *Prolegonemis Rhetoricum Hermogenis* (1570), which is not available today. Other *prolegomena* can be found in [18].

6.3.2   Leibnizian Sources for Treatment of the Case

Leibniz delves into the rhetor's case in two of his early writings. First, he discusses it
in Question 12 of the *Specimen Quaestionum Philosophicarum ex Jure Collectarum*
[13, VI 1, pp. 87–90], i.e. Leibniz's master thesis in Philosophy, which was
published 1664 in Leipzig. The second source, on which we will concentrate in
this paper, is his 1666 dissertation in law *De casibus perplexis in Jure*, in which the
case of Protagoras and Euathlus is treated in Chap. 16 [13, VI 1, pp. 241–242]. Both
works are part of his academic writings, with *De Casibus perplexis* being the last
such writing since Leibniz left the University of Altdorf to later work on a career at
the court in Mainz. The two works are similar insofar as both deal with problems that
are part of the field where law and philosophy overlap. In the case of the *Specimen,*
this is made explicitly by the very title of the work, while in *De Casibus Perplexis*
it is the subject matter, namely perplex cases, that leads to an overlap of both fields.
According to Leibniz, perplex cases are those law cases for which a solution is not
readily found within the 'normal' reaches of law. He compares them to the Gordian
knot and the Greek notion of the word *aporu*[4] (having no way through) (cf. [15,
p. 215]). His opinion on all these cases, however, is that they can still be solved
"ex mero jure" [13, A VI 1, p. 239], that is, using the tools provided by law. This
includes the case of Protagoras and Euathlus, even though for Leibniz this case falls
in a special category. Different from the ancient sources, Leibniz treats the 'puzzle'
as an actual case and attempts to find a solution in law.

6.4   The Puzzle and Its Possible Solutions

Leibniz himself lets the participants in the court case state their arguments in a short
and distinctive manner, showing their similarity as well as the argument's almost
circular nature [13, VI 1, p. 241].

The rhetor argues:

> Hac, inquit, causa, seu vinces, ex pacto mihi tenebere; seu vinceris, ex re judicata.

With either of the possible outcomes of the presentation in court, the teacher upholds
that his pupil owes him the money, the cause being either the court's decision in the
rhetor's favour, or, if the court decides in favour of the pupil, the fulfilment of the
condition set in the contract.

The pupil refutes his teacher's arguments using the same structure of argumentation:

> . . . hac, inquit, causa, seu vincam, nihil tibi ex re judicata debebo, seu vincar, nihil ex pacto.

---

[4]See Leibniz, *De Casibus Perplexis*, § 4 [13, VI I, pp. 235–236] for his explanation of the term
perplex.

The pupil claims that if he wins the court case he will have a court decision absolving him from payment, while if he loses the case, the provision in the contract will not be fulfilled, preventing his teacher from claiming the fee.

The problem becomes obvious by the diametrical opposite conclusion both parties reach: both think they will be victorious no matter the courts's decision. Leibniz thus shows that, at a first look, the judge may also be puzzled by two well-spoken rhetors providing seemingly sound arguments that are mutually exclusive, before he goes on to show that a solution is possible.

6.4.1 Traditional Solutions

In the ancient sources as well as by writers through the ages, we can find some traditional solutions to the problem. Gellius, for instance, writes that the judges refused to give a verdict; instead they delayed the decision indefinitely because they were afraid that any given verdict would contradict itself. In this version, the dilemma or paradox is seen by the judges but not solved. Different from Leibniz's view of the case, to the judges in Gellius rendition a solution using the law does not seem to be possible for this – to use Leibniz's terminology – 'perplex' case.

The result in the version attributed to Hermogenes even spawned a Greek proverb, though it can by no means be called a legal solution in the modern term. The judges listened to the arguments brought forth by both the master Corax and the pupil Tisius and became so angry with the presented hair-splitting that they threw both of them out of the court with the words: *kakoi korakes, kakon oon* (bad crows, bad egg).[5]

Many other writers, including Apuleius, give only the dilemma without any attempt at providing a solution. Those sources either want to point out the supposedly useless rhetoric stratagems used by the sophists or give the case as a starting point for the readers's own thoughts.

These sources make no effort to find a fitting verdict but dismiss the case as not really fitting a courtroom. Leibniz chooses a different approach.

6.4.2 Leibniz's Solution

Leibniz's solution to the ancient case is a legal one, taking the setting of the dilemma seriously: a court room, where any solution has to fit the legal bindings.

---

[5] A play on Corax's name, which is translated as crow/raven: from such a teacher nothing good can come. Pupil and Rhetoric are equally to blame. Cited from Leibniz [13, VI 1, pp. 241–242].

Leibniz's result seems unusual for the most quoted case in a work on perplexity: "It is our opinion that this case is incorrectly referred to as perplex".[6] When surveying the (rather limited) literature commenting on *De Casibus Perplexis* (e.g. [4, 5, pp. 198–2159, 6]), the case of Protagoras and Euathlus is quoted more often than any other example from Leibniz's work on perplexity. The case is even called "fulfilling the parameters Leibniz set for a *casus perplexus*" [4, p. 29] which is in direct opposition to what Leibniz actually wrote about the case. Leibniz shows clearly that he does not think it necessary in this case to use the tools he proposes for dealing with perplex cases.[7] Instead, 'normal' law including procedural law provides a solution when the nature of a court case is taken into account. The solution proposed is a judicial take on a case taken from an originally philosophic discussion, using Roman (procedural) law as set in the *Codex Iuris Civilis* (CIC) instead of commenting on the sophistic nature of the arguments. The answer Leibniz gives is twofold: at the time of the hearing, the judge will decide against Protagoras in favour of Euathlus. While the judge deliberates the court case, the condition set in the contract has not yet been met. Protagoras will only get his fee on the basis of the contract if Euathlus *has* won a court case.[8] A court decision can only take into account what has happened up to the pronouncement of a judgment at the latest. Up to that time, Euathlus has not won a case yet, only with the pronouncement itself the condition could possibly be met. Therefore, from a legal point of view, the judges should have no problem deciding in favour of Euathlus since the grounds for awarding Protagoras the money, the condition set in the contract, is not fulfilled yet. In Roman law asking for something more than can be awarded by law is called a *plus petitio*, as is set in C.3.10.1 f of the CIC. Here the 'asking for too much' is meant in a temporal sense, i.e. too early, before the set condition is fulfilled. This interpretation of a *plus petitio* was already accepted in classical times [11, p. 96]. The judgment in favour of Protagoras is hindered by the raised exception of the *plus petitio*. Therefore, Euathlus will win this case.

However, this is not the end for Leibniz: With the judgment denying Protagoras his money, Euathlus has now won his first case, therefore the condition of the contract is met and the situation changes. In the oldest Roman laws, if a plaintiff raised a *plus petitio* claim, everything he was owned was forfeited (see [11, pp. 337–338]). But, as Leibniz elaborates, this changed already with Zeno's decree, preserved in C.3.10.1 and amended by Justinian in C.3.10.2. Using Roman law as set in the CIC, the plaintiff would have to wait an appropriate time before returning to court. Therefore Protagoras can return to court and raise a second suit once the contract's condition (the first win) is met and he waited as long as

---

[6]*Nostra est, hunc casum immerito referri inter perplexos* [13, VI, p. 242] (author's translation).

[7]In *De casibus perplexis,* Leibniz proposes three rules to deal with perplex cases, depending on the type of perplexity (perplex disposition or perplex *concursus*) and the type of thing in dispute (divisible or indivisible).

[8]Or the first court case of his career. See [22] for the discussion on 'first win' or 'win first case' condition.

deemed appropriate by the court. In this second suit the *plus petitio* is no longer an impediment, since the exception raised is only dilatory, i.e. with the change of situation (the fulfilment of the condition) the exception is no longer valid, it only concerns the first suit and does not lead to a general loss of the case. Therefore, Protagoras will win this second case and be entitled to claim the fee from Euathlus.

Leibniz shows here that it is not necessary for the judge to give a *non liquet* judgment but a solution using basic principles of Roman law is possible. Since Roman law was still one of the main law sources in Leibniz's time, his solution seems both simple and elegant, finding a way out of the dilemma without having to recur to anything but normal procedural rules that are still reflected in most European law systems even today.[9] His view of the case could be called a 'Two Suits Solution', since it involves two separate court proceedings, necessitated by the double role the first suit fills: as a court judgment and as fulfilment of the condition set in the contract.

6.4.3   Modern-Day Solutions

During the twentieth century, quite a few writers have also attempted to solve the case of Protagoras and Euathlus, both from a legal as well as from a philosophical or logical point of view. These newer attempts at solving the dilemma should be seen in comparison to Leibniz view 300 years earlier.

One objection raised by German lawyers (cf. [12, p. 613]) is that the case would not be perplexing if raised in court today, since the German civil law system has a rule against purposefully circumventing a set condition in a contract, in § 162 I Bürgerliches Gesetzbuch (BGB).[10] According to German law, if the party benefitting from the non-fulfilment hinders the fulfilment of the condition, then a fiction of compliance comes into being and the condition is regarded as fulfilled. These authors are of the opinion that Euathlus, by not taking up any court case after having finished his training, fulfils the criteria set in § 162 I BGB. Therefore, Protagoras will win the case as the condition of the contract is deemed fulfilled.

However, there are certain objections that can be raised against this solution. For one, the German § 162 BGB requires action in bad faith on Euathlus's part (see [19, § 162, No. 9–12]). Yet, if we rely on the story as told by Gellius or Leibniz, there is no reason given as to why Euathlus did not take up any court cases. Thus, we can only speculate about Euathlus's reasons, which could have been anything

---

[9]The 'past-looking' nature of a court judgment, the type of exceptions possible and the possibility to raise a new suit once a condition is fulfilled can all be found in most law system that are at least partly based on Roman law, including Germany.

[10]A similar provision can be found in the French Code Civile, § 1178. The English translation of § 162 I BGB reads: If the satisfaction of a condition is prevented in bad faith by the party to whose disadvantage it would be, the condition is deemed to have been satisfied.

from finding no clients to actual malicious intent of not meeting the condition of his contract with Protagoras. Using German law as the context, the burden of proof for Euathlus's bad faith is resting with Protagoras, who will have trouble giving such proof, at least from the facts known about the case.

The second argument is that § 162 BGB is almost identical to Ulp. D.50.17.161. Since the German as well as the French provisions are based on this Digest rule, the solution would have probably been possible when Gellius considered the case (Ulpian also lived in the second century BC) and certainly when Leibniz formed his legal solution. As a result, this legal provision might solve the case if evidence of Euathlus's malicious circumvention of the contractual condition was available; however, such information cannot be found in the sources and thus the dilemma cannot be reliably solved this way.

Interestingly enough, one may argue that there is more ground to consider Protagoras's action as falling under Ulp. D.50.17.161 than Euathlus's. The sources provide some evidence for the fact that Protagoras brings suit against Euathlus on purpose and in order to fulfil the condition in the contract. This could be interpreted as Protagoras falling under the restriction set in Ulp. D.50.17.161 and the related modern provisions. However, taking a possible claim to court can only be considered malicious in very special cases, as it is the best available legal tool to enforce any claim. For this kind of intent on Protagoras's part we – once again – do not have enough evidence.

Another attempt at solving the case has been made by both jurists and logicians. They say the case about the contract is not the first case as referred to in the contract. Most of these attempts offer a very short explanation to their solution (cf. [21, pp. 121–122]). From a lawyer's perspective, the objection could be raised that the condition of the contract can only be met by a 'neutral' first case, not one in which the parties of the contract are disagreeing over the viability of the payment condition itself. The only good argument for that would be that both parties had not thought of this situation (the first case of Euathlus being a case about the teaching contract) when drafting the contract. One might conclude from there that this specific court case does not meet the condition set.

A possible outcome discussed by Schneider [21] is that the modern 'doctrine of frustration' could here lead to the contract being invalid. This seems to be an extreme solution and not in keeping with the potential will of the contract's parties. If they would have considered the situation (a deciding factor in the application of this doctrine), it would much more likely be their will to uphold the contract since good parts of the obligations are already fulfilled. Therefore, this solution does not seem to fit very well the requirements of the parties.

Even more unfunded seems the idea to restrict the cases 'fitting' the condition to those in which Euathlus is working for the plaintiff instead of the defendant (this is suggested but dismissed in [14, p. 166]). No evidence at all can be found in the sources to support this limitation and such a restriction seems arbitrary in light of the wording in the contract.

From a different, more mathematic perspective an objection has been raised by Northrop [17, p. 201]. Northrop mentions the story of Protagoras and Euathlus in his work "Riddles in Mathematics" as an example of a vicious circle. He attributes the difficulties to a problem of self-reference: the specific case in court concerns "all members of certain classes of things" – i.e. here all possible cases that could bring the student to court for his first case – and at the same time is a member of this class – because whether the contract is fulfilled here depends on this being the first case won, which is the property all members of the class share.

Northrop calls this vicious circle hard to avoid and suggests that a solution can be found in Russell's theory of logical types. He holds that there is a distinction between the members of a class – one type, i.e. all possible court cases that could be the 'first case' mentioned in the contract – and a statement about all the possible members of the class, which he considers as being of a different type (i.e. the contract itself). Since these are two different types, the self-reference can be avoided and there is no contradiction, or, in Leibniz's words, perplexity.

Problematic here is whether the distinction of different types is useful when applied to a legal situation. How could the outcome of the case change? The most likely result would be that the mentioned suit is not a fulfilment of the condition, i.e. that this type is excluded as an exception. This solution seems similar to the jurist's attempt by Schneider [21] mentioned above. Once again, to exclude this law suit as a 'first suit' under the condition seems hard to accept in a legal sense since a major concern for the fulfilment of a condition and the interpretation of the contract is the will of the parties entering the contract. Here there is no evidence that they wanted to exclude this situation, especially since Leibniz's 'Two Suits Solution' shows that regular juridical proceedings are sufficient to find a legal solution.

The most interesting attempts at a solution in modern times may be relying on formal logic to analyze and solve the dilemma. The authors put the arguments raised by the parties into formal language and try to arrive at a conclusion by using these formulas.

There is more than one approach using formal logic: from the very basic provided by Goosens [8, pp. 67–75] to the more complex ones developed by Aquist [3, pp. 73–84] and Lenzen [14, pp. 164–168].

The more detailed attempts show very well the opposite conclusions the parties draw from using very similar sounding arguments by formalizing the parties's arguments. From there, the authors draw conclusions and derive new statements, using formal language for all of it. These formalizations lead to a conclusion very similar to the one Leibniz arrives at using law principles: the 'perplexity' can be avoided if a temporal element is taken into consideration. That leads them to differentiate between the situations before and after the verdict is given in the (first) case. While *before* the verdict Protagoras has no claim, and thus the judges will have to rule in Euathlus's favour; with the delivery of the sentence against Euathlus, the condition set in the contract is fulfilled and therefore *afterwards* Euathlus should be successful if he brings a second case to trial.

These solutions using formal logic visualize very well the steps necessary to find a way out of the 'paradox', i.e. the introduction of a temporal element. Remarkable is that both Aquist and Lenzen arrived at the 'Two Suits Solution' without having read Leibniz.[11]

Using a second case to get out of the dilemma could be criticized as avoiding rather than solving the 'paradox' since it is not a strict philosophical approach. However, if the story is set in a court situation, it seems more than sensible to take into account the nature of a court judgment, here in its twofold effect: first of all, only using the events that already happened, not the future effects on the contract. And secondly, that here the court's decision changes the state of the condition set in the contract since with the judgment's pronunciation the contract's condition is fulfilled and Protagoras therefore has a justified claim. Both parts are not specialized procedural law of one country but rather basic rules that are used in all Roman-based law systems.

From this perspective, it may be interesting to consider Brewer's [7, p. 120] criticism of Leibniz solution as merely "legal-contingent", i.e. dependent on the applicable (procedural) law. He raises the objection that under contemporary US law the second case of Protagoras would be dismissed because of the doctrine of 'res iudicata'. The case is already decided with the first suit and a new ruling is precluded to guarantee legal stability. However, I would argue that when taking Leibniz's solution seriously, the second case is based on a change in the factual situation: only after the verdict in the first case is given, the condition set in the contract is fulfilled. Based on this new fact, a second suit must be possible: *res iudicata* only precludes a decision on the same matter if no relevant changes have happened, at least using a German understanding of this principle [2, § 322, No. 147–154]. Another 'legal-contingent' objection Brewer raises against Leibniz's solution concerns the ability of modern courts to nullify the contract in question if they found Euathlus or Protagoras to have acted in bad faith. As I noted above whether Euathlus was obliged by the contract to take cases to court or not is a factual question that would be considered in court; the same holds for Protagoras actions of suing Euathlus. However, from the information given about the case, we cannot decidedly answer these claims without further evidence, as I argued above.

I understand Leibniz solution as a logical sound one that uses the tools law provides. It is strong exactly because it achieves a solution that is usable in court and not just in theory while still solving the logical problems. Leibniz's primary claim is that this is not a perplex case, implying that 'normal' legal rules are enough to find a sound solution. It is true that in order for Leibniz's idea to be applicable, this court has to have certain procedural laws. But I assume that Leibniz only wanted to cover these circumstances. Especially in *De casibus perplexis*, a solution outside of law would go against the general thesis Leibniz set, i.e. to solve all cases within the constraints of law.

---

[11]As they state in their forward to the second issue (Aquist) and postscript (Lenzen), Leibniz's solution was pointed out to them later on.

6.5   Conclusions

Leibniz once again sheds new light on an old problem with his work on the rhetor's case. He approaches the puzzle from a new direction and attempts a 'juridical' solution. Compared to both the ancient solutions and modern attempts, Leibniz's strategy seems very plausible, provided one wants to engage with the dilemma as a serious logical and juridical puzzle and not just as an expression of criticism of Sophistic rhetoric. His unique view of problems on the border between philosophy and law gives great insights and offers especially to modern-day law scholars a good example on how using a structured argumentation and a clear understanding of the underlying principles in law can help solve problems that otherwise seem hopelessly 'entangled'.[12] "Puzzle" might be the best expression for this case from Leibniz perspective, since the emphasis in his solution is on the fact that there is no perplexity. In Protagoras's case, nothing more is needed than 'regular' law to find a satisfying or 'right' solution, i.e. a logical sound one.

References

1. J. Adamietz (ed.), *M. Fabi Quintiliani. Institutionis Oratoriae, Liber III, Part 1* (Wilhelm Fink Verlag, München, 1966)
2. J. Adolphsen et al., *Münchner Kommentar zur ZPO*, 4th edn. (Verlag C.H. Beck, München, 2013)
3. L. Aquist, The Protagoras case: An exercise in elementary logic for lawyers, in *Time, Law, and Society*, ed. by J. Bjarup, M. Blegvad (Franz Steiner Verlag, Stuttgart, 1995), pp. 73–84
4. R. Backhaus, *Casus Perplexus* (C.H. Beck'sch Verlagsbuchhandlung, München, 1981)
5. H. Ben-Menahem, Leibniz on hard cases. Archiv für Rechts- und Sozialphilosophie **79**, 198–215 (1993)
6. P. Boucher, *Des Cas Perplexes en Droit* (Librairie Philosophique J. Vrin, Paris, 2009)
7. S. Brewer, Law, logic and Leibniz. A contemporary perspective, in *Leibniz: Logico-Philosophical Puzzles in the Law. Philosophical Questions and Perplexing Cases in the Law*, ed. by A. Artosi, B. Pieri, G. Sartor (Springer, Dordrecht, 2013), pp. 99–126
8. W.K. Goosens, Euathlus and Protagoras. Logique et Analyse **77/80**, 67–75 (1977)
9. S.J. Harrison, J.L. Hilton, V. Hunink (eds.), *Apuleius: Rhetorical Works* (Oxford University Press, Oxford, 2001)
10. L. Holford-Strevens, *Aulus Gellius, an Antonin Scholar and His Achievement* (Oxford University Press, Oxford, 2003)
11. M. Kaser, *Das Römisches Privatrecht, Zweiter Abschnitt* (C.H. Beck'sch Verlagsbuchhandlung, München, 1975)
12. R. Knütel, Zur sog. Erfüllungs- und Nichterfüllungfiktion bei der Bedingung. Juristische Blätter **98**, 613–626 (1976)
13. G.W. Leibniz, Sämtliche Schriften und Briefe. Edited by the Deutsche Akademie der Wissenschaften zu Berlin. Darmstadt, 1923 sqq., Leipzig, 1938 sqq., Berlin, 1950 sqq. Cited by Series (Reihe) and Volume (Band)

---

[12]"Durcheinander wickeln" – another (German) expression Leibniz uses to describe perplexity ([13, VI, p. 236]).

14. W. Lenzen, Protagoras contra Euathlus. Betrachtungen zu einer sogenannten Paradoxie. Ration **19**, 164–168 (1977)
15. H.G. Liddle, R. Scott, *A Greek-English Lexicon* (Clarendon, Oxford, 1996)
16. P.K. Marshall (ed.), *Aulus Gellii. Noctes Atticae, Tomus I* (Oxford University Press, Oxford, 1968)
17. E.P. Northrop, *Riddles in Mathematics* (Rogert E. Krieger Publishing, Huntington, 1975) (reprint of the 1944 edition)
18. H. Rabe, *Prolegomenon Sylloge* (Teubner, Stuttgart/Leipzig, 1995) (reprint of the 1931 edition)
19. K. Rebmann, R. Rixecker, F.J. Säcker, *Münchner Kommentar zum BGB*, 6th edn. (Verlag C.H. Beck, München, 2012)
20. J.C. Rolfe, *The Attic Nights of Aulus Gellius* (Harvard University Press, London/Cambridge, MA, 1961) (reprint of the 1946 version)
21. E. Schneider, *Logik für Juristen*, 6th edn. (Verlag Franz Vahlen, München, 2006)
22. J.H. Soebel, The law student and his teacher. Theoria **53**, 1–18 (1987)
23. S. Wilcox, Corax and the Prolegomena. Am. J. Philol. **64**, 1–23 (1943)

Part III
Current Interactions Between Law and Logic

Chapter 7
On Hypothetical Judgements and Leibniz's Notion of Conditional Right

Shahid Rahman

Abstract Sébastien Magnier provides a remarkable analysis of the notion of conditional right with the help of public announcement logic that he generalizes for the logical study of legal norms. Magnier's main idea, motivated by the earlier exhaustive textual and systematic work of Matthias Armgardt and the subsequent studies carried out by Alexandre Thiercelin, involves Leibniz's notion of *certification,* which plays a central role in the famous *De conditionibus.* Magnier proposes to render the notion of *certification of A* as *there is public evidence for A.* More generally, the meanings of "conditional right" and "conditional legal norm" are established by means of identifying a specific kind of dialogical interaction during a legal trial constituted by games of giving and asking for reasons. This yields a theory of meaning rooted in the practice itself of legal debates.

The main aim of this paper is to study the notion of conditional right by means of constructive type theory (CTT) which provides the means to develop a system of *contentual* inferences rather than of syntactic derivations. Moreover, in line with Armgardt, I will first study the general notion of dependence as triggered by hypotheticals and then the logical structure of dependence specific to conditional right. I will develop this idea in a dialogical framework where the distinction between *play-object* and *strategy-object* leads to the further distinction between two basic kinds of pieces of evidence and where meanings is constituted by the interaction of obligations and entitlements.

S. Rahman (✉)
Département de philosophie, UFR Humanités, UMR 8163 Savoirs, Textes, Langage, Université de Lille 3, Villeneuve-d'Ascq, France
e-mail: shahid.rahman@univ-lille3.fr

© Springer International Publishing Switzerland 2015   109
M. Armgardt et al. (eds.), *Past and Present Interactions in Legal Reasoning and Logic,* Logic, Argumentation & Reasoning 7, DOI 10.1007/978-3-319-16021-4_7

7.1 Introduction

Sébastien Magnier [1] provides a remarkable analysis of the notion of conditional right,[1] which he generalizes for the logical study of legal norms. Magnier's main idea, motivated by the earlier and exhaustive textual and systematic work of Matthias Armgardt [2–4][2] and the subsequent studies carried out by Alexandre Thiercelin [5–7],[3] involves Leibniz's notion of *certification,* which plays a central role in the famous *De conditionibus.* According to Magnier, the *certification* of the antecedent of a sentence expressing a conditional right – such as in *If a ship arrives, Primus must pay 100 dinar to Secundus* – is linked to an epistemic understanding of evidence. In our example, the *certification of the arrival of a ship* amounts to *there is public evidence for the arrival of a ship* and this amounts to *being in possession of the knowledge required to produce a piece of evidence for the arrival of a ship.* Moreover, inspired by Kelsen's conception of legal norms, Magnier generalizes his own approach in which he rejects a material-implication approach[4] and reconstructs conditional right and legal norms in the frame of a dialogical formulation of dynamic epistemic logic that includes sentences where a *public announcement operator* occurs. In other words, Magnier's contribution consists in a shift in perspective focussing on the semantics of truth-dependence underlying the meaning of conditional rights. The main idea is to identify the epistemic dynamics involved in the fulfilment of the condition as constituting the core of the meaning of dependence specific to the notion of conditional right. He implements this shift by means of a dynamic epistemic logic called *Public Announcement Logic* (PAL).

The dialogical framework provides a further development of this dynamic by furnishing a dynamic theory of meaning. In a nutshell, the meaning of *If a ship arrives, Primus must pay 100 dinar to Secundus* boils down to establishing the conditions of a legal debate where *Secundus* claims the 100 dinar, given that the arrival of a ship has been *certified* (i.e., given that *it is known that a ship arrived,* or given that *there is evidence for the arrival of a ship*), rather than rendering this meaning by means of a model-theoretic semantics. More generally, the meaning of the notions of conditional right and legal norm is established by identifying the main logical features of those argumentative interactions that are deployed in legal trials. This leads Magnier to design specific logical language games (dialogues) that yield a theory of meaning rooted in legal practice itself.

I certainly endorse the idea that (i) a theory of meaning involving legal reasoning should be based on an argumentative-based semantics, (ii) an epistemic approach to

---

[1]In the present paper the term is used in the sense of Leibniz rather than in the sense in which it is generally understood in legal contexts nowadays.

[2]The work of Matthias Armgardt prompted and influenced a host of new research on the bearing of Leibniz's approach to current studies in legal rationality.

[3]In fact, Thiercelin's research was prompted by the work of Armgardt.

[4]In fact, Magnier [1, pp. 151–157; 261–292] rejects other forms of implication interpretations too, including strict implication or connexive implication.

the notion of *legal evidence* should have a central role in a theory of legal reasoning, and (iii) implication is not really at stake in the logical analysis of conditional right. However, I think that the role of evidence should be given prominence and developed into a general epistemic theory of meaning where evidence is understood as an object that makes a proposition true. More precisely, I think that we should explore the possibility of placing the piece of evidence that grounds a proposition (the object that makes the proposition true) at the object-language level, instead of via the formal semantics of an operator that introduces that evidence via the metalogical definitions of a formal (model-theoretical) semantics. That a proposition is true is supported by a piece of evidence, but this piece of evidence must be placed in the object language if that language is purported to have content. This move seems to be particularly important in the context of legal trials where acceptance or rejection of legal evidence is as much part of the debate as the main thesis itself. More generally, the notion of legal evidence should be linked to the meaning of a proposition and not only of an operator occurring within a proposition.

The main aim of the present paper is to study the notion of conditional right by means of a constructive type theory (CTT) according to which propositions are sets, and proofs are elements. That a proposition is true means the set has at least one element. The analysis of legal norms should follow as a generalization, the details of which are not the subject of the present paper. In such a framework, the logical structure of sentences expressing conditional rights is analyzed as corresponding to that of hypotheticals rather than implications. The proof-objects that make the implications of the hypothetical true are *pieces of evidence dependent upon the evidence for the condition* (i.e. *dependent upon the evidence for the head of the hypothetical*). Herewith I follow Thiercelin's [6, 7] interpretation that considers the notion of dependence as the most salient logical characteristic of Leibniz's approach to conditional right. Moreover, in line with Armgardt [2, pp. 220–225], I will study the general notion of dependence as triggered by hypotheticals and then the logical structure of dependence specific to conditional right. However, in my view, the dependence of the conditioned on the condition is defined with regard to the pieces of evidence that support the truth of the hypothetical rather than the propositions that constitute it. According to this analysis, the famous example for a conditional right:

If a ship arrives, then Primus must pay 100 dinar to Secundus

has the form of the hypothetical

Primus must pay 100 dinar to Secundus, provided there is some evidence x for the arrival of a ship

And this means

The evidence *p* for a payment-obligation that instantiates the proposition *Primus must pay 100 dinar to Secundus* is dependent on some evidence *x* for a ship arrival

Furthermore, the *general logical structure* of the underlying notion of dependence yields:

$$p(x) : P \ (x : S)$$

where x is a yet unknown element of the set of arrivals S (i.e. $x : S$), and where the evidence for a payment-obligation (the piece of writing that establishes the conditional right) is dependent on the arrival x of a ship, i.e., the evidence for payment-obligation is represented by the function $p(x)$.

In this setting, when there is knowledge of some ship arrival s, the variable will be substituted by s.

Still, the logical structure $p(x) : P (x : S)$ represents the more general case of dependence triggered by an underlying hypothetical form which is common to all right-entitlements that are dependent upon a proviso clause – such as the *requirements clause* of statutory right-entitlements or the *condition clause* of conditional right-entitlements. Moreover, a further deeper analysis requires an existential quantification embedded in a hypothetical of the sort:

If $(\exists w : S)$ *Arrive(w)true, then Pay* $(100\ dinar,\ primus,\ Secundus)$ *true*

Even this deeper analysis does not seem to fully capture the *future contingency* of the conditions upon which conditional rights are built.[5] Nevertheless, this formalization $p(x) : P (x : S)$ provides a general formal approach to the notion of dependence that, as pointed out by Armgardt [2, pp. 221–225], seems to be in line with Leibniz's [8, VI, I, p. 235] own approach to the generalization of right-entitlements by means of hypotheticals.

As regards the specificity of conditional right, Leibniz himself defended, on one hand, a biconditional reading of the notion of dependence,[6] and on the other hand, the uncertainty regarding the fulfilment of the condition at the moment of the formulation of a (legally valid) concrete case of conditional right-entitlement.[7]

If we consider explicitly the underlying epistemic and temporal structure in the way that Granström [9, pp. 167–170] tackles (in the CTT-frame) the issue on future contingents, a biconditional formalization specific to *Leibniz's notion of* condition-dependence is possible. As a matter of fact, Aristotle's chapter of the *Peri Hermeneias* on the sea battle naturally leads to Leibniz's example of the ship. Roughly, the underlying idea is that both implications hold:

> If a ship arrives then, Primus must pay 100 dinar to Secundus, (provided (S or not S) and assuming that the arrival of a ship proves the disjunction).
> If Primus must pay 100 dinar to Secundus, (provided (S or not S) and assuming that the arrival of a ship proves the disjunction), then a ship arrival is the case.

---

[5]This was suggested by Göran Sundholm in a personal email.

[6]The biconditional reading relates to the link between the condition and the conditioned. Leibniz calls this feature of the conditional right *convertibility*. It is not clear if, in Leibniz's view, the biconditional reading only applies to conditional right.

[7]This seems to be rooted in actual legal practice: If the condition A is not satisfied, the benefactor is not entitled to B. The actuality of this feature of the Leibnizian approach to the notion of conditional right has been defended by modern-day scholars of Law theory such as Koch / Rüßmann [10, p. 47] and more thoroughly by Armgardt [2–4].

However, it seems that a general approach does not require biconditionality after all, at least not in its full extension. Regarding the link between condition and conditioned, it only requires a hypothetical conjunction constituted by the following implications:

> If the condition C is fulfilled then the beneficiary is entitled to the right at stake, assuming that some evidence for C solves the uncertainty (C or not C) underlying the conditional right.
>
> If the condition not C is fulfilled then the beneficiary is not entitled to the right at stake, assuming that some evidence for not C solves the uncertainty (C or not C) underlying the conditional right.

Furthermore, I will develop this idea in a dialogical framework where the distinction between *play-object* and *strategy-object* (*or proof-object*) leads to the further distinction between two basic kinds of pieces of evidence such that strategy-evidence is made up of play-evidence. The proposed approach includes the study of *formation rules* that model the argumentation on the acceptance of a piece of evidence.

I do not claim to have captured all the complex issues related to the notion of legal evidence, but the aim is to explicate more precisely the logical and semantic place it should occupy in legal reasoning in general and in conditional right in particular.

The present paper is divided into two main parts, with two appendices in which the main features of the formal background of parts I and II are presented.

In the first part, entitled **Leibniz's Logical Analysis of the Notion of Conditional Right and Beyond**, I propose a study of the notion of conditional right by means of the CTT approach to hypotheticals based on Leibniz's logical analysis.

In the second part, entitled **Dialogical Logic and Conditional Right**, I develop the analysis of the previous section in an appropriate dialogical framework. In so doing, I adopt Magnier's idea that the pragmatic semantics of dialogical logic can capture (some of) the properties that Leibniz ascribes to a notion of conditional right rooted in actual legal practice.

Appendix I: The basic CTT-frame for intuitionistic predicate logic
Appendix II: Linking Dialogical Logic and CTT

7.2   Leibniz's Logical Analysis of the Notion of Conditional Right and Beyond

At the early age of 19, Leibniz embarked upon a logical analysis of the notion of conditional right that provided insights which still inspire modern-day researchers in the field of legal reasoning. The main aim of the present chapter is to revisit those insights with a view to gaining new perspectives for the current understanding of the logical structures underlying the legal meaning of the notion of conditional right. In that sense, I follow the positions taken, on the one hand, by Magnier, whose remarkable 2013 study elucidates the role of epistemic evidence in Leibniz's analysis, and on the other hand, by Thiercelin [5–7], whose interpretation – based on the earlier textual and systematic work by Matthias Armgardt [2–4] – considers the notion of dependence as the most salient logical characteristic of Leibniz's approach to conditional right. However, I will depart from Armgardt's, Thiercelin's and Magnier's approach to the extent that I will develop my proposal in the framework of a CTT-formulation of hypotheticals.

7.2.1   Leibniz's Logical Analysis of Conditional Right

During the period 1664–1669 the young Gottfried Wilhelm Leibniz (1646–1716) studied the theory of law with the prolific creativity that was to make his fame. It is during this period that Leibniz developed his theory of conditional right in two main texts that provided the content for two academic dissertations:

1. *Disputatio Juridica (prior) De Conditionibus* [8, VI, I, pp. 97–150], which was defended in July and August 1665. At that time Leibniz was a 19-year-old student who had already received the title of Master of Philosophy in February 1663 following the defence of his *Disputatio Metaphysica De Principio Individui* in December 1662.
2. *Disputatio Juridica (posterior)De Conditionibus* [8, VI, I, pp. 97–150], which is part of Leibniz's *Specimina Juris* (1667–1669),

A modified version of his theory is given in *Specimen Certitudinis Seu Demonstrationum In Jure, Exhibitum In Doctrina Conditionum* [8, VI, I, pp. 367–430], which is part of Leibniz's *Specimina Juris* (1667–1669), a compilation and reformulation of three of his already held disputations: the *Disputatio Inauguralis De Casibus Perplexis In Jure*, that granted Leibniz the doctoral degree in November 1666. The prior and subsequent disputations constitute the main source of the present discussion.

7.2.1.1   Suspension as Dependence

As pointed out by Thiercelin [5–7], the main point of the young Leibniz's work on conditional right is to provide a logical analysis of the juridical modality known as *suspension* (the term was first used by Roman jurists), which should stress the specificity of conditional right in relation to other conditional propositions such as those of geometry or those expressing causal necessity. The novelty of Leibniz's approach, which was developed as an answer topuzzles raised by Roman jurists,[8] is

---

[8]Pol Boucher [11] provides a thorough discussion on the Roman sources of Leibniz' own developments.

that it considers suspension as affecting the condition of a conditional proposition. According to Leibniz, the notion of condition (and its modality), as it pertains to the study of conditional right, should be studied in the context of its role in affecting the truth of a *proposition* that expresses some legal right provided by an individual agent (the benefactor or *arbiter*) in favour of a second individual (the beneficiary). The effect of the (suspensive) condition δ of a conditional right is that its beneficiary is entitled to a certain right if the condition δ (established by the benefactor) is fulfilled.

The approach underlying Leibniz's proposal is that the role of the notion of condition specific to conditional right is that of introducing a dependence relation such that the *truth of the proposition* that expresses the conditioned is said to be dependent upon the *truth of the condition*. This, in Leibniz's view, and from a logical point of view, is what *suspension* is about: the truth of a proposition is *dependent* on the truth of a given condition established by the arbiter (benefactor).

Thus, Leibniz's analysis of the notion of conditional right is based on a logical study of propositions, and consequently, as thoroughly discussed by Armgardt [2–4], the ancient links between logic and law are implemented in a novel way (see also [12]). In fact, Leibniz [8 VI, I, p. 101] searches for a logical system that makes legal reasoning *almost as certain as that of mathematical demonstrations*. Now, the logical form of a proposition that best accommodates this analysis is that of the conditional sentence that expresses some specific type of hypothetical. Hypotheticals that formalize a conditional right are constituted by an antecedent that Leibniz [8, VI, I, p. 235] calls the *fact* and a consequent that he calls *jus*. Thus, Leibniz's main claim is that hypotheticals such as

If a ship arrives from Asia, then Primus must pay 100 dinar to Secundus

provide an appropriate approach to the meaning of juridical formulations such as

Secundus's right to receive 100 dinar from Primus is suspended until a ship arrives from Asia.

However, as already mentioned, Leibniz wished to distinguish those hypotheticals that formalize conditional rights – he calls them *moral conditionals* - from other forms of hypotheticals that share some logical and semantic properties with moral conditionals.[9] In order to do so Leibniz fixes logical, epistemic and pragmatic properties that should characterize moral conditionals. Suspension also has epistemic and pragmatic features specific to the legal meaning of moral conditionals. Let us briefly discuss each of these separately, though, as we will see below, these levels are interwoven in crucial ways.

---

[9]See Vargas [13] for a thorough discussion on Leibniz' view on the links between moral conditionals and hypotheticals of geometry and causal necessity.

Truth-Dependence and Convertibility

In Leibniz's view the main logical property of moral conditionals is, as already mentioned above, that of the *dependence* of the truth of the **jus** (the consequent of the hypothetical) on the **fact** (the antecedent of the hypothetical). Now, since this dependence is established by the will of the *arbiter* (the benefactor) it also has a pragmatic (Leibniz calls it *moral*) feature. The pragmatic outcome of the creation of such a form of dependence is that if the condition is not fulfilled there is no ground for the legal claim.[10]

Now, since, according to Leibniz's view, the logical understanding of dependence amounts to the truth-dependence of the **jus** on the **fact**, Leibniz concludes that the logical structure of moral conditionals is such that if the antecedent of the hypothetical is false, then so is its consequence. However, since he also takes contraposition to be one of the axioms that characterize moral conditionals, formally speaking, the notion of dependence leads him to concede that moral conditions are – in their pure logical form – biconditionals. Leibniz [8, VI, I, p. 375] summarizes this by saying that the condition and the conditioned of moral conditionals are *convertible*.[11] Our author certainly sees that convertibility, from the pure logical viewpoint, might blur the crucial difference between condition (**fact**) and conditioned (**jus**). Leibniz's [8, VI, I, p. 112] strategy out of the dilemma is to point out that, (i) though formally speaking the antecedent and the consequent of a moral conditional are convertible, from a legal point of view the pair **fact-jus** is analogous to the pair **cause-effect**, and (ii) though in hypotheticals expressing a relation of cause and effect, the antecedent and consequent *are accomplished together*, the *condition starts to exist first*. Note that, of its own, the analogy threatens to undermine the claim that convertibility is the specific property of moral condition. A way to further develop Leibniz's response is to stress that, according to this approach, the logical convertibility of moral conditions is the effect of a specific act of will that provides the hypothetical with legal meaning. Thus, if we were to adopt this viewpoint, we should claim that what makes some hypotheticals moral conditions is that the legal meaning of the underlying conditional right grounds the dependence between condition and conditioned. Hence, the difference between hypotheticals that express a **cause and effect** and hypotheticals that express conditional right is to be found in the meaning on the basis of which the respective dependencies are defined: while **cause and effect** dependence is defined on the basis some notion of natural necessity, **fact-jus** dependence is defined on the basis of the will of the arbiter in such a way that the dependence of the **jus** upon the **fact** is the result of legal acts acknowledged as such by competent authorities. Thus, in general, it is not the dependence itself but rather the meaning of the notion of dependence involved that distinguishes moral conditionals from other hypotheticals. Following such a path requires a thorough

---

[10]Cfr. Leibniz [8, VI, I, p. 375]. For a discussion on this issue see Thiercelin [7, p. 207].

[11]For a further discussion on the criticism of the biconditional rendering of the suspensive modality, see Magnier [1, pp. 155–156].

description of how meaning triggers the targeted notion of dependence. Armgardt [2, pp. 362–363] points out that Leibniz's logic of legal reasoning underlies an (incipient) *Conceptography*. Unfortunately, Leibniz does not develop– at least not explicitly – the link between the notion of dependence and the logic of concepts. Nevertheless, it might argued that Leibniz's argument on the interrelation between condition and conditioned mentioned above delivers the elements for linking the logical structure of conditional right with the logic of concepts intrinsic to this notion.

Be that as it may, Pol Boucher [11] seems to think that though in the context of legal reasoning Leibniz continues to be a rationalist who looks for general patterns of inference, he adopts here some sort of Gricean procedure. According to this interpretation, dependence is a logical property –manifested by convertibility – but the difference between condition and conditioned is a presupposition of legal practice. In fact, it seems that this coincides with standard legal practice even nowadays. This practice seems to serve as the basis of Koch / Rüßmann's defence [10, p. 47] of the biconditional reading of the dependence of the **jus** upon the **fact**, and of Armgardt's [2] subsequent careful study. In fact, as mentioned in the introduction, and as I will discuss in Sect. 7.3.2 below, a sophisticated form of Leibniz's take on biconditionality can be worked out that articulates the interaction between meaning and logic features as discussed above. This seems to relate to Marcelo Dascal's [14, 15] findings of a *soft rationality* in Leibniz work. According to my view on soft rationality, Dascal's interpretation suggests that, in the context of legal reasoning, it is crucial to see that the notion of rationality behind this kind of reasoning is the result of the interaction between meaning and logic features, or more generally, between syntax, semantic, pragmatic and epistemic features – and this connects with Armgardt's remark on the interaction between legal reasoning and Conceptography in Leibniz's work on the logic of Law.

Alexandre Thiercelin [5–7] proposes to tackle the issue on convertibility by means of connexive logic. The proposal is sensible since connexive logic has its roots in the Stoic tradition that Leibniz was certainly familiar with. The axiom of connexive logic relevant to the formalization of moral conditionals is:

$$(A \Rightarrow B) \Rightarrow \neg (\neg A \Rightarrow B) \text{ (where " } \Rightarrow \text{ "is the connexive conditional)}$$

This is quite similar to convertibility, but without falling into the total convertibility of condition and conditioned.[12] Indeed, the point is that the moral condition *If A, then B* (connexively) implies that it is not the case that if the condition is not fulfilled the *jus* is true. Moreover, the moral conditional *If A, then B* also implies that it is not the case that if the condition is fulfilled the *jus* is false:

---

[12]Rahman and Rückert [16, 17] and Rahman and Redmond [18, pp. 20–57] provided a dialogical semantics for it based on the idea that this conditional is a particular kind of strict implication (defined in S4) where the head is satisfiable and the negation of the tail is also satisfiable. See also Pizzi and Williamson [19], Priest [20], and Wansing [21, 22].

$(A \Rightarrow B) \Rightarrow \neg (A \Rightarrow \neg B)$ (where " $\Rightarrow$ "is the connexive conditional)

The problem with this conditional is that its semantics departs significantly from standard classical logic. It is not even a conservative extension of it.[13] Furthermore, as signalized by Thiercelin [7, pp. 208–11] and criticized by Magnier [1, pp. 157–159], the epistemic component has to be added in some ad hoc manner.

Let us now examine some of the epistemic features of moral conditions.

Suspension and Its Epistemic Nature

Suspension has a crucial epistemic feature – that was very well known in Roman Law and explicitly discussed by Leibniz – which has become part of the definition of conditional right in current legal systems, namely the legal validity of a concrete case of conditional-right entitlement requires a situation where the fulfilment of the condition is not yet known at the moment of its formulation. In fact, as stressed by nearly all scholars on Leibniz's work on the logic of law, one of Leibniz's main original contributions is to have linked this epistemic feature with a conditional logical structure. Some Roman jurists, for example, connected the non-fulfilment of the condition with existential issues that could account for this non-fulfilment. Leibniz's logical solution is clear and simple: suspension amounts to the dependence of the truth of the conditioned upon the truth of the condition, combined with the uncertainty about the truth value of the condition. This assumes that although the conditional right ascribes a right to the beneficiary (dependent upon the fulfilment of the condition), the truth value of the condition might not yet be known to be true. The lifting of the suspension amounts to what Leibniz [8, VI, I, p. 424] calls *certification* of the fulfilment of the condition (that is, the production of an item of evidence for the fulfilment of the condition). Moreover, as I will discuss in Sect. 7.3.2, Leibniz's solution can be linked to the temporal structure underlying assertions on future contingents.

Magnier' proposal [1, pp. 141–187] is based on a shift of perspective to the truth-dependence underlying conditional rights. The main idea is to identify the epistemic dynamics involved in the fulfilment of the condition as constituting the core of the meaning of dependence specific to the notion of conditional right. Thus, in Magnier's view, truth-dependency is a consequence of the epistemic nature of suspension. Magnier implements this shift by means of the use of *Public Announcement Logic* (PAL). Indeed, the PAL-approach to conditional right allows Magnier to describe how the initial model (defined for a PAL sentence expressing a given conditional right), that might include scenarios where the condition is fulfilled and scenarios where it is not, changes when it is known that the condition

---

[13]Note that, according to Rahman and Rückert´s semantics, although $A \Rightarrow A$ is connexively valid, the implication $(A \Rightarrow A) \Rightarrow (A \Rightarrow A)$ is not. For further non-classical features of their semantics (see [17, pp. 106–108; 120–121]).

is fulfilled (the initial model shrinks to a model that only contains scenarios that fulfil the condition).[14] Furthermore, Magnier's work not only provides a new formal framework for capturing the dynamics inherent to the logical structure of the notion of conditional right, it also highlights, for the first time, as far as I know, another aspect of the epistemic nature of the condition attaching to a conditional right, namely that the *certification* of the fulfilment of the condition must be the object of public knowledge.[15] This public aspect of the epistemic nature of the condition was inspired by Kelsen, and the PAL-reconstruction allows Magnier to express in the same framework all the epistemic features, namely:

1. the uncertainty regarding the fulfilment of the condition at the moment of the formulation of a concrete case of (legally valid) conditional-right entitlement;
2. the epistemic dynamics triggered after the fulfilment of the condition; the dynamics also concern the temporal dimension of the notion of suspension signalized by Armgardt [2, pp. 349–351];
3. the requirement of public knowledge of the evidence for such a fulfilment.

Magnier's approach also can also deal with the truth dependence in way that involves some subtle distinctions between a *false announcement* regarding the fulfilment of the condition, and the assertion that the condition is false. In fact, in the PAL framework, there is no way to express a false announcement at the object language level. Leibniz's certification of the condition corresponds to asserting it, such that if we certify that *A is false* we assert that *non-A is true*. And this is certainly different from making a false announcement. If there is a false announcement, the epistemic updating process is, so to say, aborted and hence the truth-value of the whole PAL sentence cannot be established. From the viewpoint of legal practice, a false announcement corresponds to making it public that the condition has been fulfilled while it has not, and hence, presumably, either we are forced back to the initial situation where the condition has not yet been fulfilled, or the whole obligation expressed by the conditional right is declared to be null and void.[16]

---

[14]$[\varphi]\psi$ *is true at the evaluation world* **s** *iff* φ *is true at* **s** *implies that* ψ *is true at the reduced model* $\mathbf{M}\big|\varphi$ – *where the reduced model* $\mathbf{M}\big|\varphi$ *is the result of removing from* **M** *all the worlds where* φ *is false:* $\mathbf{M}, s1 = [\varphi]\psi$ iff $\mathbf{M}, s1 = \varphi$ *implies* $\mathbf{M}\big|\varphi, s1 = \psi$.

[15]In fact, this epistemic requirement for the fulfilment of the condition was already pointed out by Thiercelin [6, p. 141]. However, it was not incorporated in the logical analysis of the conditional before Magnier's work.

[16]The case of cancellation corresponds to that of PAL: since false announcements cannot be made, i.e. the system aborts. The first case, where a false announcement forces us back to the initial models, corresponds to the *Total Public Announcement Logic* (TPAL) of Steiner and Studer [23], where false announcements do not change the initial model at all. Indeed, if it is taken that it has been announced that *A* is true, then if we come to know that that *A* was not true after all, the original PAL process stops there since the model shrinks and there is no way of going back. However, one can imagine a system, such as TPAL, that allows us, once we come to know that *A* is not true, to go back to the model before it shrinks, since the grounds adduced for reducing it in the first place are no longer available. This might also relate to Armgardt's recent work [24] on the defeasibility

Certainly, this is different from certifying the falsity of the condition. In this case it is the truth-value of the tail (Magnier calls it the *post-condition*) of the PAL sentence that will not follow. The dialogical game of the certification of the falsity of the condition shows that the Proponent will win, but he will win his thesis about the truth of the whole PAL sentence without engaging at all on any assertion involving the post-condition. Independently of the distinctions discussed above, in such a framework, if the PAL-sentence is true, then it cannot happen that the post-condition B will be evaluated as true, even though the condition A cannot be announced (because A is false). Taken together, the PAL approach to conditional right yields the following rendering of the truth-dependence:

The condition is true iff the PAL sentence expressing the conditional right is true.

Still, there are some arguments drawn from legal practice for the biconditional reading of dependence. The point is that if the condition is false, then the claim for the right involved in the conditional right at stake will be rejected. If we follow Magnier's approach the analysis will yield the following: Since in this framework it cannot happen that the post-condition can be evaluated as true without the condition being true, and the jury must evaluate the post-condition as true in order to ascribe the right claimed by the benefactor. Hence, in the case that the condition is false, the jury will not be able to evaluate the post condition as true (there will be no grounds to support the post-condition) and will reject the claim. However, as pointed out before, the falsity of the condition A[17] does not entail that $[A]B$ is false. This corresponds to some cases of legal practice where although the beneficiary might not be entitled to the claimed right, this does not mean that the conditional right is not legally valid.

Perhaps, if we wish to continue along the PAL path to conditional right after all, we might formulate its logical form as the conjunction:

$$[S] \, O \, P \wedge [\neg S] \neg O \, P$$

or

$$[S] \, O \, K \, P \wedge [\neg S] \, K \neg O \, P$$

(or some other combination of modal operators in the tail of the PAL-sentence)

Note that $[S]OP$ and $[\neg S] \neg O$ are not contradictory: the submodel for the left PAL-sentence contains worlds were S is true, and the submodel for the right PAL-sentence excludes those worlds where S is true, so, after the update, both submodels will contain different worlds, and in none of them will we have S and not S.

---

of the grounds adduced for the establishment of a fact (in our case the defeasibility of the grounds adduced for the fulfilment of condition).

[17]Recall that the falsity of the condition is different to producing a false announcement.

Now, besides the logical and epistemic features underlying the structure of the notion of conditional right, there are also pragmatic aspects that contribute to the, so to say, moral aspect of the suspensive modality.

Suspension and Its Pragmatics

Conditional rights are structures with legal content. It is the content that interacts with some features of the underlying logical and epistemic structure. As discussed above, epistemic features are essential to the (legal) definition of conditional right. But this content and the validity of concrete conditional-right entitlement are also determined by pragmatic features that qualify conditional right as conditional and not as some other kind of right-entitlement under assumption. The most decisive of these pragmatic features is that which determines that the attribution of a conditional right to some beneficiary is due to the sole will of a benefactor (and not of the legislator). This assumes that the arbiter should be factually and legally able to ascribe the conditional right at stake, and that condition and conditioned meet specific legal requirements. Sometimes the underlying legal meaning of the conditional right hinges on the envisaged target of the arbiter regarding the fulfilment of the condition. The ultimate goal of engaging in a particular conditional right might be directly dependent on the arbiter's interest in motivating the beneficiary to fulfil a given condition. The pragmatic features also interact with the logical structure of the conditional-right entitlement attributed to a given benefactor. For example, legal systems will rule out impossible or unlawful conditions, and similarly for the **jus** part of a conditional right. Hence, it seems sensible to require that both the condition and the conditioned are logically and factually possible. It does not make any legal sense to formulate a conditional right involving a logical truth or a logical falsity.

Thiercelin [7, p. 213] remarks that scholars in the field do not seem to have paid very much attention to these practical aspects of the notion of conditional right, despite the fact that Leibniz [8, VI, I, p. 409, 422] himself makes a careful study of the cases of what Roman jurists called *ridiculous* conditions. In my view, the point here, once again, is the interaction of content with logical structure. However, the standard-model-theoretical approach to semantics places this interaction at the metalogical level. A clear example of this metalogical viewpoint is the semantics of PAL deployed by Magnier, where the whole epistemic dynamics manifests itself in the formal semantics of the model, and the latter is metalogically defined. What we need is a language where we can check, at the object-language level, the meaning of a given expression and furthermore that a given proposition is true.[18] This is linked with the formation plays mentioned above. Before we check the truth of a given proposition we need to check its meaning – its legal meaning – against the background of knowledge of the legal system. Let us explore this line of thought.

---

[18]For a thorough discussion on the criticism to the model-theoretical approach to meaning see Sundholm [25–29].

7.2.2 Hypotheticals and Conditional Right

The following approach is based on Leibniz's idea that the most salient characteristic of the logical structure of the notion of conditional right is the truth dependence of the conditioned upon the condition. Moreover, although I will adopt Magnier's epistemic shift, I will propose a new shift that takes us from the epistemic nature of the propositions to the objects (the pieces of evidence) that ground the knowledge required by the notion of suspension.

7.2.2.1 Hypotheticals and the General Form of Dependence

In the CTT framework it is possible to express that A *is true* at the object-language level by means of the assertion *d: A (there is a piece of evidence d for A* or *there is a proof-object d for A)*.[19] Therefore, within this framework the dependence of the truth of B upon the truth of A amounts to the dependence of the proof-object of the former to the proof-object of the latter. The dependence of the proof-object of B upon the proof-object of A is expressed by means of the function *b(x)* (from A to B), where x is a proof-object of A and where the function *b(x)* itself constitutes the dependent proof-object of B. As discussed in Appendix A.I, dependent proof-objects provide proof-objects for hypotheticals, for instance:

$$b(x) : B \ (x : A) \, ,$$

which reads, *b(x)* is a (dependent) proof-object of B provided x is a proof-object of A.

In our context, proof-objects, in principle,[20] correspond to pieces of evidence. Thus, the dependence of the truth of the **jus** B upon the truth of the condition A boils down to the fact that the piece of evidence for B is the function *b(x)*.

It follows from this analysis that the notion of dependence relevant for Leibniz's famous example for a conditional right:

If a ship arrives, then Primus must pay 100 dinar to Secundus

can be expressed by means of the hypothetical:

Primus must pay 100 dinar to Secundus, provided there is some evidence x for the arrival of a ship

And this means

---

[19]See Appendix A.1.

[20]In Sect. 7.3, we will distinguish between play-objects and strategy-objects. While the latter correspond to proof-objects, pieces of evidence or evidence might be either play- or strategy-objects.

The evidence p for a payment-obligation that instantiates the proposition *Primus must pay 100 dinar to Secundus* is dependent on some evidence x for a ship arrival.

This would naturally lead to rendering the underlying logical structure with the help of a hypothetical, which roughly amounts to

$$p(x) : P\ (x : S)$$

where, x is a yet unknown element of the set of arrivals S (i.e. $x : S$), and where the evidence for a payment-obligation (the piece of writing that establishes the conditional right) is dependent on the arrival x of a ship, i.e., the evidence for payment-obligation is represented by the function $p(x)$.

A deeper – though not yet definitive – rendering of the logical structure is the following[21]

If $(\exists w : S)\ Arrive(w)$***true***, *then* Pay $(100\ dinar,\ primus,\ Secundus)$***true***

and this then demands a hypothetical proof

$$b(x) : Pay\ (100\ dinar,\ Primus,\ Secundus)$$

under the hypothesis that

$$x : (\exists w : S)\ Arrive(w)$$

Even this deeper analysis does not fully capture either the *future contingency* or the *convertibility* of those conditions that build conditional rights. However, as mentioned in the introduction, in the context of law, generally speaking, this logical structure is shared by all other forms of right-entitlement with proviso clauses, such as statutory-right entitlements under *requirements*. According to this analysis, all of them share some form of hypothetical structure the meaning of which is provided by dependent proof-objects. Further distinctions are necessary in order to distinguish between them. For instance, while requirements for statutory-right entitlements do not demand uncertainty about the satisfaction of these requirements, conditions of conditional-right entitlements, as discussed above, do. Furthermore, both have a different origin: while requirements are established by the legislator, conditions are established by the sole will of the arbiter.

Nevertheless, the study of the general form is desirable from both the logical and the legal point of view. Furthermore, as discussed by Armgardt [2, pp. 221–225], Leibniz [8, VI, I, p. 235] himself pointed out that hypotheticals provide the general logical form of those right-entitlements where the proviso clause (such conditions or requirements) corresponds to the *antecedent* of the hypothetical and the *consequent* to its **jus**.

---

[21]This analysis was suggested by Göran Sundholm in a personal email.

According to this analysis, a general and basic form of right-entitlement is that of an hypothetical with a proviso clause that provides the conditions/requirements under which the proposition is made true This seems to coincide with Leibniz's and current legal terminology, where a right is granted on the occurrence of *fact*. Thus, in line with this analysis, the logical structure of such kinds of right-entitlements *is not* that of a proposition but that of a hypothetical that binds assertions (or judgments)[22] in such a way that the assertion of the **jus** is made dependent on the assertion of the condition. For instance:

$$p(x) : P\ (x : S)$$

The point of the formalization is that we can formulate explicitly at the object-language level that the pieces of evidence for the fulfilment of the antecedent of the hypothetical are not yet known, namely by the use of variables. It is the variables for pieces of evidence that make right-entitlements with proviso clause hypotheticals. More precisely, in the context of CTT, the variable in a hypothetical such as $p(x) : P(x : S)$ represents *an unknown element of S* that can be instantiated by some s when the required knowledge is available.[23] Thus, in this framework, instantiating the *unknown* element x by some s *known* to be a fixed (but arbitrary) element of S is what Leibniz's notion of *lifting the suspension* is about. Using the current terminology of epistemic logic as an analogy – in the style of Hintikka [30] – where we say that a judgment of the form

$$x : S$$

expresses *belief* rather than *knowledge* and that

$$s : S$$

represents the transition from belief to knowledge, we suggest, in the context of our discussion, that this might also represent the transition from a right-entitlement under *the hypothesis (or belief) S to be case* (i.e. $x : S$) to *a right-entitlement grounded by the knowledge that the condition/requirement S has been satisfied* (i.e., $s : S$). In fact, for this transition to count as a transition to knowledge, it is not only necessary that $s : S$, but it is also necessary that it is known that the piece of evidence s (a concrete ship arrival) is the piece of evidence of the adequate sort (cfr. Ranta [31, pp. 151–154]). In other words, we also need to have the definition

$$x = s : S$$

---

[22]Recall that a judgement or assertion expresses that a proposition is true. The assertion *A is true* introduces an epistemic feature: it is known that A is the case. Furthermore, judgements can also involve sets: *1 is an element of the set of Natural numbers* (in the CTT-notation: $1 : N$).

[23]Cf. Granström [9, pp. 110–112]. In fact, chapter V of [9] contains a thorough discussion of the issue.

This definition of x can be called an *anchoring* of the hypothesis (belief) S in the *actual* world (Ranta [31, p. 152]). Thus, the result of this anchoring process yields

$$p\,(x = s) : P\,(s : S)$$

If there is more than one hypothesis (including interdependences – temporal or otherwise – between them; a requirement for statutory-right entitlements can be dependent on other requirements, and the same holds for conditions of conditional-right entitlements),[24] it is not required that all the variables will be substituted at once. It is possible to imagine a gradual reduction of uncertainty through a gradual introduction of definitions of the variables – in the case of temporal interdependences, the graduality of the fulfilment is determined by a fixed order. A general formulation of this kind of transition[25] is the following, where Γ and Δ are hypotheses that represent some kind of proviso (such as conditions or requirements) for right-entitlements:

$$\Gamma = (x_1 : A_1, \ldots x_n : A_n) \text{ becomes}$$
$$\Delta = (\Gamma, \; x_k = a : A_k)$$

such that, in the new hypotheses, every occurrence of x_k is substituted by a. The new hypothesis Δ is obtained from Γ by removing the hypothesis $x_k : A_k$ by $a(x_1 \ldots x_n)$. Thus, as required, this operation reduces the uncertainty within the original hypothesis.

Let us now examine the path that takes us from the general to the specific.

7.2.2.2   Granting Statutory and Conditional Rights

The main features that distinguishes a statutory right from a conditional right are

(a) the uncertainty concerning the fulfilment of the condition attaching to the conditional right – this does not apply to the requirements proviso of statutory rights;
(b) the (ontological) type of the individual that grants the correspondent right. While the individual that grants a conditional right is a person (natural or legal), the individual that grants a statutory right is a legislator.

---

[24] Armgardt [2, p. 256] explores different kinds of interdependences between conditions discussed by Leibniz [8, A VI I, pp. 387–388], including temporally ordered conjunctions and subsidiary ones such as *If it is known that the condition A cannot be fulfiled then the condition B should be fulfiled*. Armgardt [1] makes ample use here, and in other parts of his book, of temporal indexes.

[25] This transition is known in CTT as *definitional extension of hypotheses or contexts*. For the formal details, see the presentation in Appendix AI.2.

Let us turn our attention to (b) - we have already discussed (a) above and it will be discussed further on in the next section. The CTT framework provides the means to make the ontological type of the individual that grants a certain right explicit at the object-language level. Recall that the CTT framework claims that syntactic and meaning traits are to be processed at the same time and both of them occur at the object-language level. For instance, before proving the logical validity of a sentence, it is required to display its content, and the latter amounts to ascribing the adequate type to each part of the sentence (see AI.) – and, when appropriate, to identifying the canonical elements of the correspondent sets. In our case, let us take the following hypothetical:

$b(x)$: B $(x : A)$ – recall that $b(x)$ is the dependent object that constitutes one of the pieces of evidence for the hypothetical

For the sake of simplicity, let us for the moment ignore the inner (existentially quantified) structure of A. Let us further assume that the piece of evidence $b(c)$[26] – the contract – expresses a statutory right R^S:

$$R^S \ (b(c)) \ \ true$$

This presupposes that $R^S(y)$ is a proposition, provided $y : B$; that $b(c) : B$; and that $b(c)$ is the result of a substitution in the function $b(x)$ from A to B.[27] In other words, $(b(x) : B \ (x : A)$ and $c : A)$.[28] Moreover, the explicit presentation of the presuppositions involved requires displaying the putative pieces of evidence that might count as acceptable – this might also involve describing the *canonical elements* of the sets involved (see Appendix A.1).

Following a simplified form of legal terminology, we say that $b(c)$ is a *statutory right* iff it is granted ($G(y, w, z)$) to person y (natural or juridical) by a *legislator y*[29]:

$$R^S \ (b(c)) \ \ iff \ G \ (l, \ b(c), \ p) \ true$$

Where G (y, w, z) is a proposition, provided $y : legislator, w : B, z : person)$ and it is *known* that $l : legislator, b(c) : B$ and that $p : person$.

For the sake of clarity of exposition, I do not quantify over the function-dependent objects $b(x)$. However, a full development of the definitions of statutory and conditional right should quantify universally over them.

[27]For the formation rules for functions, see Appendix A.1.

[28]For further details on the formation rules involved, see Appendix A.1.

[29]In fact, as pointed out by Armgardt in a personal email, we need the following parameters: Who grants What, Whom, When, and based on Which legal norm. I did not add all of the parameters to avoid a heavy notation. Let me mention that in the CTT-frame the introduction of temporal indexes in the object language is pretty straight-forward – see Ranta [31, pp. 101–124].

This, in turn, presupposes that the instance x of A and w of B are *neither illegal nor against boni mores* $(\neg M(w))$. Since w is some function $b(x)$ defined on A, it seems to be sufficient to require this restriction on the function only:

$$M \; (b(c)) \;\; true$$

Where $M(w)$ is a proposition, provided $w : B$, and it is known that $b(c) : B$.

In fact, from the point of view of law, granting occurs independently of knowing whether or not the proviso has been satisfied. Thus, we need to delve deeper into the structure of the proviso in order to achieve generality:

$$B \;\; true \;\;\; ((\exists v : V) \; A(v)) \, true$$
$$b(x) : B, \text{provided} \;\; x : ((\exists v : V) \; A(v))$$

This yields the general form of a grant

$$R^S \; (b(x)) \;\; iff \; G \, (l, \; b(x), \; p) \, true$$

Thus, the following should be included in the list of presuppositions:

$$b \;\; \text{is a function from} \;\; x : ((\exists v : V) \; A(v)) \, \text{to} \;\; B.$$

The explicit presentation of presuppositions by means of formation rules seems to be very natural to a legal trial. This is one of the main motivations for the use of a dialogical frame (see Sect. 7.3 below).

Similarly, let us now say that $b(c)$ expresses an instance of a *conditional right* iff it is granted $(G(\dots))$ to a person y (natural or juridical) by a *person z*. Indeed, the difference between a statutory and a conditional right is, in this respect, the type of the individual that provides the grant: a *person* in the latter case and a *legislator* in the former.

A last tricky point concerns the *closing* of the engagement expressed by the statutory/conditional right once the proviso has been fulfilled by *one* instance. I will come back to this issue in Sect. 7.3. Let us first examine the logical form of conditional rights.

7.2.2.3 The Specificity of Conditional Rights: Uncertainty with and Without Biconditionals

One of the most difficult issues regarding the logical structure of conditional rights relates to the fact that the obligation expressed by the **jus** is made dependent on the occurrence of a future, uncertain *event*. In other words, on the occurrence of a future contingent event – the example of the ship recalls almost explicitly Aristotle's sea-battle case. On the other hand, as discussed above, Leibniz's approach and current legal practice seem to lead to the idea that convertibility is at the core of the logical form of conditional rights. In the context, once more, of the ship example: If we know that a ship has not arrived, it seems that it should be inferred that it is not the case that *Primus* must pay. The point here is to find a formalization that makes these two crucial features of conditional rights explicit.

However, it seems that, although the logical form of conditional rights requires that if it is known that the condition will never be fulfilled, then the right-entitlement should fail, it *does not require* that if the beneficiary is entitled to the right involved, the condition has been fulfilled. In fact, in section "Uncertainty and Convertibility Without Biconditionals" we claim that biconditionality is not necessary after all.

The Logical Form of Leibniz's Approach: Uncertainty and the Biconditional

In order to implement this double task, I will provide the head of the hypothetical with a richer structure than the one discussed above. More specifically, I take it that the head of hypotheticals underlying conditional right have the form of a constructivist disjunction. That is, a disjunction, such that proof-object of it, amounts to indicating explicitly which of both obtains. Thus the head looks like:

$$x : S \vee \neg S$$

(there is some piece of evidence x for the disjunction: *a ship arrival is the case or not*)

Since we are in the framework of a constructive disjunction, its truth requires that we know which of both obtains. In our case, we need to know that that ship did arrive. Let us express this with

$$(\exists y : S) \ (y = x)$$

(there is some ship arrival and this ship arrival constitutes the evidence for the disjunction – i.e., it is equal to the evidence x for the disjunction $S \vee \neg S$)

However, the above notation does not explicitly express that y constitutes evidence for the left part of the disjunction. *It is precisely the left part of the disjunction that represents the condition required by the conditional right of our example.* In order to do so we need to make use of the function *left(y)*. This yields:

$$(\exists y : S) \ (left(y) = x)$$

Clearly, the arrival of a ship (which constitutes the fulfilment of the condition) implies that *Primus* must pay. That is,

$$a(x) : (\exists y : S) \ (left(y) = x) \rightarrow P \ (x : S \vee \neg S)$$

Moreover, since we postulate that *Primus* must pay if the disjunction is true, and since we established that the disjunction is true if a ship arrives (and not the contrary), it follows that a payment obligation dependent on the disjunction implies the arrival of a ship:

$$b(x) : P \ (x : S \vee \neg S) \rightarrow (\exists y : S) \ (left(y) = x)$$

If we pull it all together we obtain what could be taken as the logical form underlying *Leibniz's notion of conditional obligation:*

$$(\exists y : S) \ (left \ (y) = x) \leftrightarrow P \ (x : S \vee \neg S)$$

The proof-object of which is an object dependent upon x:

$$d(x) : (\exists y : S) \ (left(y) = x) \leftrightarrow P \ (x : S \vee \neg S)$$

However, x is the evidence for the fulfilment of the condition. Thus

$$d \ (left(y)) : (\exists y : S) \ (left(y) = x) \leftrightarrow P \ (x : S \vee \neg S)$$

Still, this seems to express the biconditionality of the condition in relation to the whole hypothetical. So can we also prove the *convertibility* of condition and conditioned? Indeed, this is the case, as I show in the following paragraph.

Let us show that if the condition is false, then so is the conditioned. That is, let us show that if $\neg S$ *true*, it is also the case that $\neg P$ *true*. Let us assume that we have a dependent proof-object for the biconditional:

$$d(x) : (\exists y : S) \ (left(y) = x) \leftrightarrow P \ (x : S \vee \neg S)$$

If $\neg S$ *true*, then the evidence x for the disjunction consists in some z : $\neg S$, such that x is *right(z)*. Thus, by substitution we have:

$$d \ (right(z)) : (\exists y : S) \ (left(y) = right(z)) \leftrightarrow P \ (x : S \vee \neg S)$$

The left part of the biconditional is clearly false, and thus so is the right part, and hence $\neg P$ *true*.

Thus, if $\neg S$ *true*, it is also the case that $\neg P$ *true*, as required.

Uncertainty and Convertibility Without Biconditionals

Let us assume that there have been two conditional agreements: One makes the obligation to pay 100 dinar dependent upon the arrival of aship, and the other,

dependent on another condition, say, the arrival of a caravan from Asia. In such
a case, the biconditional seems to be too strong a requirement since it might be
that we know that no ship will ever arrive. However, this does not mean that
Secundus has no payment obligation. Leibniz's discussion on disjunctive conditions
[8, VI, I, p. 388] might provide a kind of solution. From this viewpoint, the logical
structure of the condition is in fact a disjunction and this, as pointed out by Armgardt
[2, pp. 267–269], can be embedded in a biconditional structure. The problem with
this solution is that every conditional-right agreement must be completed with
perhaps not yet known disjunctive elements such as *If a ship or a caravan or ...*
or ... arrives, then P. Moreover, if there are more agreements, it is not established
whether we have to add the conditions of the different agreements as disjunctions
or conjunctions, or if we have to interpret the different agreements as different
conditional-right entitlements. In the latter case, *Secundus* might be entitled to 200
dinar: 100 when a ship arrives, other 100 when a caravan does. Nowadays lawyers
tend to follow Leibniz – who was an active lawyer – and switch to a pragmatic
strategy. Confronted with such cases, the court decides in view of the best possible
interpretation of the benefactor's will.

Another possibility is to give up the (full) biconditional structure, and propose
the following hypothetical conjunction:

$$d(x) : (((\exists y : S) \ (left(y) = x)) \rightarrow P) \wedge$$
$$((\exists z : \neg S) \ (right(z) = x)) \rightarrow \neg P)) \ (x : S \vee \neg S)$$

Which reads:

> *If there is some evidence for a ship arrival, and this arrival solves the uncertainty (S or not
> S) underlying the conditional right, i.e., if the ship arrival provides evidence for the **left side**
> of the disjunction, then the beneficiary is entitled to the right at stake.*
>
> *If there is some evidence for no ship arrival and this solves the uncertainty (S or not S)
> underlying the conditional right, i.e., if the evidence for no ship arrival provides evidence
> for the **right side** of the disjunction, then the beneficiary is not entitled to the right at stake.*

In fact, this (hypothetical) conjunction of implications seems to be the most
suitable formalization of the logical and epistemic structure underlying the notion
of conditional right.

Indeed, such a conjunction ensures that *if it is known that a ship will never
arrive, there is no obligation to pay*. But what will be blocked is that *If Primus
is obliged to pay, then a ship did arrive*. Clearly, *Primus*, might be obliged to
pay because a caravan arrived, even if a ship did not. This approach still leaves
it open if *Primus* must pay 100 dinar twice if both a caravan and a ship arrived.
However, it is precisely the blocked implication that renders justice to the possibility
of interpreting two agreements (with the same **jus**, but with two different conditions)
as two different conditional entitlements or not, without assuming some tacit or
retroactive enrichment of the logical form involved. Certainly,by contraposition on

the implication of the right side of the conjunction, we obtain that $\neg\neg P \rightarrow \neg\neg S$. However, this is not a full biconditional. It only says that if it is impossible that a payment obligation is not due, then it is impossible that a ship arrival did not occur.

Interpreting Conditions

In legal practice, interpretation is crucial. Not only in order to apply the general norm to a particular case (see end of Sect. 7.2.2.1) but also in order to complete or further elucidate the terms of a given formulation. In the case of conditional right, the interpretation processes based on the logical structure of the underlying hypothetical are of three kinds:

1. application to a particular case;
2. extension of the set of conditions, interdependent or not (this includes interdependencies induced by temporal order);
3. specifying the structure of the evidence required to fulfil the condition

As the first two cases have been already discussed at the end of Sect. 7.2.2.1, let us turn our attention to the third case. Let us assume that the formulation of the conditional right is somewhat vague regarding the condition to the fulfilled in order to obtain B. To take a simple case:

$$d : B \, (u : A\vee \sim A)$$

The interpretation process will consist here in an extension to the new context, for example:

$$B \, (x : \neg A)$$

by means of a mapping, such that:

$$P \, (u = right(x) : A \vee \neg A)$$

This provides the information required to formulate the logical form of the conditional right precisely:

$$d(u) : (((\exists y : A) \, (left(y) = x)) \rightarrow \neg B) \wedge$$
$$((\exists z : \neg A) \, (right(z) = x)) \rightarrow B \,)) \, (x : A \vee \neg A)$$

In fact, the third case generalizes the others. Indeed, as explained in Appendix AI.2, these kinds of interpretation processes consist in a mapping that extends the original context into a new one with less uncertainty (cfr. Granström [9], Chap. V).

7.3   Dialogues, Play-Objects, and the Dynamics
of Conditional-Right Entitlements

Besides the general aim of developing a pragmatic semantics for legal reasoning
rooted in its specific argumentative practices, the dialogical setting provides insights
into the dynamics underlying the meaning of the notion of conditional right. Indeed,
the language games typical of dialogical logic – the dialogues – distinguish the
play level from the *strategy level*,[30] and this distinction, as discussed below, makes
it possible to study the dynamics of a trial involving a particular instance of
conditional right.

During a trial, there are moves which are purely logical and others which
are not. Among the latter, there are moves concerning the legal validity of the
original conditional-right entitlement. For instance, the validity of a given contract
involves questions of content. There may also be moves that question some pieces
of evidence. And lastly, there may be moves concerning the closing of a trial in light
of the submitted evidence.

Each of these points is developed in the following sections. However, they will
be preceded by a general presentation of dialogical play-objects for conditional
right. Note that the dialogical plays are meant to work as language games, and
are not descriptions of actual practices. In other words, they are purported to
be, in Wittgenstein's words, measurement rods, constructions by means of which
understanding and insight might be gathered. It is worth noting that, to explain this
point, Wittgenstein used the reconstruction of a fact in a trial.

7.3.1   Dialogical Play-objects as Language Games for Trials
on Conditional Right

The following form of play is the simplest one and only involves the entitlement
claim of the *ship example* once the legality of the contract and the piece of
evidence for the fulfilment of the condition (the ship arrival *s*) have been accepted.[31]
Moreover, the premise will not be the one with the conditional form but the
one that does not assume convertibility. However, in the context of how the play
described below is being developed, the difference between the conjunctive and
the biconditional form is irrelevant. Indeed, in the play to be developed below, the
Proponent chooses the side of the conjunction that involves the implication from
the non-negative condition to the conditioned, and this implication occurs in both
Leibniz's biconditional sentence and in the one without it. Thus, the premises are:

---

[30]See Appendix AII.4

[31]The formation play will not be developed here (see Appendix A.II).

(I)

$$d(x) : (((\exists y : S)\,(left(y) = x)) \to P) \wedge ((\exists z : \neg S)\,(right(z) = x)) \to \neg P\,))$$

$$(x : S \vee \neg S)$$

(II)

$$s : S.$$

The proponent, *Secundus*, claims that, grounded on I and on the piece of evidence *s* of a ship arrival S (II), he is entitled to the payment *P* of 100 dinar by *Primus*. Thus, the thesis is:

$$R^{\to}(a(s)) : P$$

Notation

Recall that *S* stands for the set of ship arrivals, *P* for *Primus must pay 100 dinar to Secundus*, *s* for the arrival of the ship *s*, and *a(s)* a concrete payment obligation, grounded on the ship arrival *s*.

In Dialogical Logic *left(x)* is written *L(x)*. In order to differentiate between the left side of a disjunction and a conjunction, exponentials will be added. However, as already discussed in A. II, the CTT-operator *left(x)* and the Dialogical instruction $L^{\vee}(x)$) have quite different roles. Indeed, the CTT-operator indicates that *x* is a proof object for the corresponding disjunction – namely a proof object for the left side of the disjunction. In contrast, given a play object *x* for the disjunction – composed by two play objects such that each of them constitutes a *sufficient* play object for the disjunction – the expression $L^{\vee}(x)$ instructs the *defender* to choose the play object for the left side of the disjunction. The same applies to difference between the expression *right(x)* and $R^{\vee}(x)$. Accordingly, the main sentence is rewritten as (Table 7.1):

$$d\left(L^{\vee}(x) : (((\exists y : S)\,y = L^{\vee}(x)) \to P) \wedge \right.$$
$$((\exists z : \neg S)\,z = R^{\vee}(x)) \to \neg P\,))\ \ (x : S \vee \neg S)$$

Decoding Keys

Move 1: After setting the thesis and establishing the repetition ranks[32] **O** launches an attack on the thesis with his first move asking for the play-object for *P*.

Move 2: P posits $p : S \vee \neg S$ in order to require the substitution *p* for *x* (see posit substitution in AII).

---

[32]Repetition ranks establish how many times a player can defend or challenge the same move (see Appendix A.II). The notion of *rank* plays an important role in modelling the closing of trials – see 7.3.3 below.

Table 7.1 A play for a claim based on a conditional right contract

	O		P		
I	$!d(L^{\vee}(x)) : \big(((\exists y : S)\, y = L^{\vee}(x)) \to P\big) \wedge \big(((\exists z : \neg S)\, z = R^{\vee}(x)) \to \neg P\big)\ (x : S \vee \neg S)$		$R^{\to}(a(s)) : P$		0
II	$!s : S$				0
II	$n := 1$	I	$m := 2$		
1	$R^{\to}(a(s)) = ?$	0	$p : P[\text{LAST MOVE: P WINS}]$		28
7	$!d(L^{\vee}(p)) : \big(((\exists y : S)\, y = L^{\vee}(p)) \to P\big) \wedge \big(((\exists z : \neg S)\, z = R^{\vee}(p)) \to \neg P\big)\ (p : S \vee \neg S)$	1	$!r : S \vdash \neg S\ (/x)$		2
3	$?\,\boxed{S \vdash \neg S}$	2	$!L^{\vee}(p) : S$		4
5	$L^{\vee}(p) = ?$	4	$!s : S$		6
9	$!L^{\wedge}(d(L^{\vee}(p))) : ((\exists y : S)\, y = L^{\vee}(p)) \to P$	7	$?[((\exists y : S)\, y = L^{\vee}(p)) \to P]$		8
11	$!a(L^{\vee}(p)) : ((\exists y : S)\, y = L^{\vee}(p)) \to P$	9	$L^{\wedge} = ?$		10
13	$!a(s) : (\exists y : S)\, y = s \to P$	11	$s/L^{\vee}(p)$		12
25	$R^{\to}(a(s)) : P$	13	$!L^{\to}(a(s)) : (\exists y : S)\,(y = s)$		14
15	$?^{\exists}_{L}$		$!L^{\exists}(L^{\to}(a(s))) : s : S$	14	16
17	$?^{\exists}_{R}$		$!R^{\exists}(L^{\to}(a(s))) : s : S$	14	18
19	$L^{\exists}(L^{\to}(a(s))) = ?$		$!s : S$	16	20
21	$R^{\exists}(L^{\to}(a(s))) = ?$		$!s : s = s\ (in\ S)$	18	24
23	$!s : s = s\ (in\ S)$		$?\ s\text{-Id-refl}$	II	22
27	$r : P$		$R^{\to}(a(s)) = ?$	25	26

Moves 3, 4: O decides to counterattack, before applying the required substitution. **P** defends with the left side.

Moves 5, 6: O asks for the play-object of the instruction $L^\vee(p)$. For its defence **P**, utters $s : S$ - **P** is allowed to do so, even if $s : S$ is an elementary posit since $s : S$ is the premise II.

Move 7: After **P**'s move 10, **O** has no other option than to carry out the substitution.

Move 8, 9: P challenges the conjunction by asking for the left side. **O** answers immediately

Moves 10, 11: P asks for the substitution term for the instruction $L^\wedge$ and **O** chooses $a(L^\vee(p))$

Move 12: **P** asks for the substitution $s / L^\vee(p)$. According to the substitution-rules, if an instruction has been already substituted this substitution once it a player can ask to carry it out in any further occurrence of the same instruction. In our case, **P** can ask **O** to substitute $L^\vee(p)$ for s since $L^\vee(p)$ has been already substituted in move 6 with s and this instruction occurs in **O**'s move 11. From the point of view of CTT this amounts recognizing that both object are equal elements in S.

Move 14: **P** challenges the implication uttered by **O** at 13.

Moves 15, 16, 17, 18: **O** decides to launch a counterattack on the existential before defending: he asks for the first member of the existential, then the second. **P** chooses s as the substitution term for y as an answer to the first challenge and accordingly carries out the same substitution for the defence of the second side of the existential.

Moves 19, 20: **O** asks for the play-objects terms of the embedded instructions $L^\exists\left(L^\to (a(s))\right)$. In other words, **O** asks for the play-object that corresponds to the left side of the existential occurring in the left side of the conditional. **P** defends by positing that the required play-object is s and this yields the posit $s : S$.

Moves 21–24: **O** asks for the play-objects of the embedded instructions $R^\exists\left(L^\to (a(s))\right)$. In other words, **O** asks for the play-object that corresponds to the right side of the existential occurring in the left side of the conditional. This forces **P** to utter $s : s = s$ but **O** has not yet posited this elementary sentence. Thus, **P** launches a counterattack on the premise II, on the grounds of the reflexivity rule for the equality predicate (see Appendix AII.5.2) and forces **O** to posit the required reflexivity. This allows P to posit $s : s = s$ at move 24.

Move 25: **O** posits $R^\to (a(s)) : P$ as an answer to **P**'s challenge launched with his move 14.

Moves 26: **P** asks for the play-objects of the embedded instructions $R^\to (a(s)) =$. In other words, **P** asks for the play-object that corresponds to the right side of the conditional.

Moves 27: At last **O** is forced to utter $r : P$.

Move 30: **P** posits $r : P$ which is exactly the same as **O**'s 29th move. This is in fact the last move of the play and the Proponent – the beneficiary's side - wins!

The reader might want to check that if the premise is that a ship arrival is not the case, a dual play can be developed in favour of the benefactor. The point is now to make use of the right side of the initial conjunction (premise I).

Note that this is still not a winning strategy. A winning strategy would involve showing that the series of moves of this play is one of the terminal series that will always lead to a win (for the notion of strategy, see Appendix AI.2). Now, certainly during a play, as in legal trials, "silly" or "logically not optimal" moves are always possible. At move 18 for example, the Proponent might have had chosen b as a substitution for y, i.e., for whatever reason, the Proponent might have put forward a ship arrival different to s as evidence. Logically, this is a weak move since both the antagonists have already agreed that s has been *certified* (to use Leibniz's words). New evidence introduced during the play is totally procedural and might be contested. This brings us to the next section.

7.3.2   Formation Plays and How to Challenge Evidence

The specific content of a given instance of a conditional right definitely concerns the development of *formation* plays where the legality of the terms constituting the contract can be questioned. The development of these kinds of formation plays involves displaying not only the elements of its logical structure but also the elements required by its legal validity, such as who granted the conditional right to whom, when, and if it is clear that it fulfils the requisite of not being either illegal or against *boni mores* (see Sect. 7.2.2.2 above). Now, it might be the case that the elements mentioned above might be introduced *during* a trial, and they might be contested on the spot. However, although a strategic viewpoint requires an overview of all the possible plays (see Appendix AII.4), this does not seem to be either necessary or desirable.

In fact, Magnier [1] in several parts of his book, in the context of a dialogical reconstruction of conditional right, discusses the case where a proposition has been certified and introduced, so to say, *for the sake of the discussion*. Accordingly, he distinguishes between two types of plays: those that are formal (purely procedural) and those that are not formal (material). If we apply this very useful distinction for the play-objects that provide the evidence for elementary posits, we can distinguish between *procedural* and *non procedural* evidence. The latter amounts to assuming that one player introduces new evidence that was not discussed or agreed upon at the start of the play. In such a case, a new formation play might start by asking for its *exact typing*, that is, for the description of the type it belongs to – roughly, for the description of the fact or proposition it is purported to support. In general, we might indeed distinguish between both kinds of evidence. If the typing is correct and legally valid, the antagonist might accept it, for the sake of the discussion. However, the antagonist might reject it and demand an examination of the new evidence. If the result of such an examination is not clearly cut it might lead to the introduction of *presumptions* with formulations such as: *it is presumed, in the absence of evidence to the contrary, that the evidence a provided a suitable play-object for B.* Moreover, perhaps in practice, the force and suitability of every piece of evidence is the result of a presumption. In such a case, the difference between purely procedural and non-procedural might lose its clear edges and we might require some kind of non-monotonic approach – if we are willing to go down that path. But the place to develop such an approach should start at the level of the formation plays. Certainly, formation plays and their correspondent main plays are intimately linked: if a piece of evidence is contested in the middle of a play, the following logical moves might require some revision. This is where the recent works of Armgardt [24, 32] on Presumptions, of Gabbay and Woods [33] on the relevance of pieces of evidence, and of Magnier [1] on the burden of the proof can be linked with the framework developed in the present paper. However, once more, this is facilitated by the play level, where pieces of evidence might not be considered indisputable facts, and therefore might be rejected. I cannot develop this link here, but this is part of a future discussion where the recent work of Giuseppe Primiero [34, 35] on Belief-Revision in the context of CTT seems to provide a useful tool. However, let me point out that the formation plays involved might involve a rich structure: they might involve issues of legality – see Sect. 7.2.2.2. The present dialogical approach to CTT provides the semantics place where such content-based challenges can be examined and developed.

7.3.3 Dialogues and the Closing of Conditional Right-Entitlements

Let us assume that one piece of evidence has been rejected – as in the preceding paragraph – and that confronted with the production of a new piece of evidence a new trial starts again. However, once the condition has been fulfilled and the claim has been judged as legally valid the process is closed. No new piece of evidence will – under normal circumstances – entitle the beneficiary once more to a new claim in the context of the same conditional-right contract at stake. For instance, returning once more to the ship: If a ship has arrived and it has been decided that the 100 dinar must be paid to *Secundus*, in general, a new arrival will not entitle *Secundus* to a further 100 dinar. The win of a play by the beneficiary, who made use, during the play, of a particular instance of the condition (a particular ship arrival), may not win again, with a new arrival, on the sole grounds that the condition is existentially quantified.

Also in this case, it is the play level that provides the right insight. Plays might be classified by their *repetition rank*. As presented in Appendix AII, repetition ranks establish how many times a player can defend or challenge the same formula. Let us assume that the Proponent agreed that, for an existentially quantified condition, one instance is sufficient. To make it more concrete, assume that the benefactor, *Primus*, agreed that one ship arrival is sufficient for the fulfilment of the condition that entitles *Secundus* to a payment obligation. Within such a repetition rank (namely 1) there might be a logically infinite number of different plays, each of them satisfying the condition with a different ship arrival. But the point is that the closing of a trial on such a right-entitlement is defined on a play!

The closing of a trial is modelled by the end of a given play within the context of a fixed rank. Moreover, if the play is lost by the benefactor-party because, for example, the purported piece of evidence was not such, this does not preclude that another trial can be run when a new piece of evidence is brought forward.

7.4   Conclusion

The central problem with the use of subjectively grounded opinions is their impotence when drawn into disagreements. Nothing is advanced in our disagreement by my putting forward opinions that you reject (cfr. [36, pp. 325–326]). This certainly applies to legal debates and it seems to be at the root of Leibniz's interest in the logic of Law and his efforts towards a unification of the fields involved. As pointed out on several occasions by John Woods, one of the best unification results of interdisciplinary work is one in which the offspring does not owe its identity to any one of its parents, though it certainly carries the marks of them: *It produces results that are very much worth having and which neither parent is able to deliver by its own* (cfr. Gabbay and Woods [37, p. 196]).

The dialogic approach is born of the idea that the rational way of overcoming disagreements is by the means of an interactive understanding of reasoning and meaning. According to the dialogical approach, meaning is constituted by and within interaction. Now, where the interdisciplinary amity of law and dialogical constructive logic is concerned, it is perhaps still too early to make a definitive assessment regarding the achievement of the kind of unification described by Woods. However, it seems safe to be fairly optimistic. The very point is that if the field of legal reasoning is to have its own identity, this must be based on an approach to meaning that provides insights into the structure of legal argumentative practices. The Dialogical approach to Constructive Type Logic provides a setting where the legal content shapes the resulting formal system in some specific ways. The proposed unifying dialogical setting harbours further features that also constitute new challenges, namely, the study of direct and indirect evidence and the epistemic dynamics specific to the legal notion of evidence. This strongly suggests that the work on presumptions mentioned above is one of the tasks to be tackled next.

Acknowledgments Many thanks to Matthias Armgardt, Nicolas Clerbout, Johan Granström and Göran Sundholm, who were closely involved with the writing of the present text from its conception down to the nitty-gritty details of the last stages. The inputs of Granström and Clerbout reflect the results of ongoing joint papers, and those of Sundholm and Armgardt further research collaborations. Moreover, without the inputs on CTT by Granström and Sundholm, and the ones on the theory and history of Law by Armgardt, the paper would not have been possible. In fact, the paper is the first result of 1 year of research in the framework of the Franco-German ANR-

DFG project, JURILOG (ANR11 FRAL 003 01), between the Universities of Konstanz (Chair of Civil Law and History of Law, Prof. Dr. Matthias Armgardt) and Lille (Chair of Logic and Epistemology, Prof. Dr. Shahid Rahman), supported by the *Maison Européenne des Sciences de l' Homme et de la Société* - USR 3185. I would also like to thank the following permanent members of JURILOG for fruitful discussions and remarks: Giuliano Bacigalupo (Konstanz), P. Canivez (Lille3), Sandrine Chassagnard-Pinet (Lille2), Karl-Heinz Hülser (Konstanz), Bettine Jankowski (Konstanz), Sébastien Magnier (Lille3), Juliette Sénéchal (Lille2) and Juliele Sievers (Lille3). The paper is also linked with my research within the "Argumentation, Decision, Action" (ADA) research program supported by USR 3185 mentioned above and the "Langage, Argumentation et Cognition dans les Traditions Orales" (LACTO) networks.

A.1 Appendix I: The Basic CTT-Frame for Intuitionistic Predicate Logic

A.1.1 AI. 1. CTT and Intuitionistic Logic

The following presentation and (brief) commentary of the rules has been extracted literally from the (more exhaustive) text of Martin-Löf [38, pp. 13–15 and 20–21] with the only minimal notational variation of using ":" instead of "ϵ". We present the rules for *product* Π (the applications of which include universal quantification and material implication), *disjoint union* Σ (the applications of which include existential quantification, subset separation and conjunction) and for *sum* (or *coproduct*) $+$ (the applications of which include disjunction).

Given a set A and a family of sets $B(x)$ over the set A, we can form the *product*:

Π **-formation**

$$
\begin{array}{ll}
\quad\quad (x : A) & \quad\quad (x : A) \\
A \; set \quad\quad B(x) \; set & A = c \quad\quad B(x) = D(x) \\
\hline
(\Pi x : A) \; B(x) \; set & (\Pi x : A) \; B(x) = (\Pi x : C) \; D(x)
\end{array}
$$

Π **-introduction**

$$
\begin{array}{ll}
\quad\quad (x : A) & \quad\quad (x : A) \\
\quad\quad b(x) : B(x) & \quad\quad b(x) = d(x) : B(x) \\
\hline
(\lambda x) \, b(x) : \; (\Pi x : A) \; B(x) & (\lambda x) \, b(x) = (\lambda x) \, d(x) : \; (\Pi x : A) \; B(x)
\end{array}
$$

Note that these rules introduce canonical elements and equal canonical elements, even if $b(a)$ is not a canonical element of $B(a)$ for $a : A$. Also, we assume that the usual variable restriction is met, i.e. that x does not appear free in any assumption except (those of the form) $x : A$.

Π -elimination

$$c : (\Pi x : A) \ B(x) \ a : A \quad c = d : (\Pi x : A) \quad B(x) \ a = b : A$$

$$Ap \ (c, a) : B(a) \qquad Ap \ (c, a) = Ap \ (d, b) : B(a)$$

We have to explain the meaning of the new constant Ap (Ap for Application). $Ap(c;$ $a)$ is a method of obtaining a canonical element of $B(a)$, and we now explain how to execute it. We know that $c : (\Pi x : A)B(x)$, that is, that c is a method which yields a canonical element $(\lambda x) \ b(x)$ of $(\Pi x : A)B(x)$ as result. Now take $a : A$ and substitute it for x in $b(x)$. Then $b(a) : B(a)$. Calculating $b(a)$, we obtain as result a canonical element of $B(a)$, as required.

The second group of rules is about the *disjoint union* of a family of sets.

Σ -formation

$$\begin{array}{cc} & (x : A) \\ A \ set & B(x) \ set \end{array}$$

$$(\Sigma x : A) \ B(x) set$$

Σ -introduction

$$\begin{array}{cccc} & & (x : A) \\ a : A & b : B(a) & A = C & B(x) = D(x) \end{array}$$

$$(a, b) : (\Sigma x : A) \ B(x) \qquad (\Sigma x : A) B(x) = (\Sigma x : C) \ D(x)$$

In fact, any canonical element of $(\Sigma x : A)B(x)$ is of the form (a, b) with $a : A$ and $b : B(a)$ by Σ−introduction. But then we also have $a : C$ and $b : D(a)$ by equality of sets and substitution. Hence $(a, b) : (\Sigma x : C)D(x)$ by Σ-introduction. The other direction is similar.

Σ -elimination

$$\left(\begin{array}{c} x : A, y : B(x) \\ c : (\Sigma x : A) \ B(x) \qquad d \ (x, y) : C \ ((x, y)) \end{array} \right.$$

$$\left. E \ (c, (x, y) \ d \ (x, y)) : C(c) \right)$$

where we presuppose the premise $C(z) \ set \ (z : (Sx : A)B(x))$, although it is not written out explicitly. (To be precise, we should also write out the premises $A \ set$ and $B(x) \ set \ (x : A)$.) We explain the rule of Σ-elimination by showing how the new constant E operates on its arguments. So assume we know the premises. Then we execute $E(c; (x; y)d(x; y))$ as follows. First execute c, which yields a canonical element of the form (a, b) with $a : A$ and $b \ 2 \ B(a)$. Now substitute a and b for x and y, respectively, in the right premise, obtaining $d(a; b) : C((a; b))$. Executing $d(a, b)$ we obtain a canonical element e of C((a; b)). We now want to show that e is also a canonical element of $C(c)$. It is a general fact that, if $a : A$ and a has value b, then $a = b : A$ (note, however, that this does not mean that $a = b : A$ is necessarily formally derivable by some particular set of formal rules). In our case, $c = (a;b) : (x : A) B(x)$ and hence, by substitution, $C(c) = C((a;b))$. Remembering what it means for two sets to be equal, we conclude from the fact that e is a canonical element of $C((a; b))$ that e is also a canonical element of $C(c)$.

We now give the rules for the *sum* (*disjoint union* or *coproduct*) of two sets.

$+$−**formation**

$$\frac{A \ set \quad B \ set}{A + B \ set}$$

The canonical elements of $A + B$ are formed using:

$+$−**introduction**

$$\frac{a : A \qquad b : B}{i(a) : A + B \quad j(b) : A + B}$$

where i and j are two new primitive constants; their use is to give the information that an element of $A+B$ comes from A or B, and which of the two is the case. It goes without saying that we also have the rules of $+$−introduction for equal elements:

$$\frac{a = c : A \qquad b = d : B}{i(a) = i(c) : A + B \quad j(b) = j(d) : A + B}$$

Since an arbitrary element c of $A + B$ yields a canonical element of the form $i(a)$ or $j(b)$, knowing $c : A + B$ means that we also can determine from which of the two sets A and B the element c comes.

$+$−**elimination**

$$\frac{\begin{array}{cc}(x : A) & (y : B)\end{array}}{c : A + B \quad d(x) : C \ (i(x)) \quad e(y) : C \ (j(y))}$$
$$\frac{}{D \ (c, (x)d(x), (y)e(y)) : C(c)}$$

where the premises A *set*, B *set* and $C(z)$ *set* $(z : A + B)$ are presupposed, although not explicitly written out.

+−equality

$$
\begin{array}{c}
(x : A) \qquad\quad (y : B) \\
a : A \qquad d(x) : C\,(i(x)) \qquad e(y) : C\,(j(y)) \\
\hline
D\,(i, (a), (x)\ d(x), (y)e(y)) = d(a) : C\,(i(a))
\end{array}
$$

$$
\begin{array}{c}
(x : A) \qquad (y : B) \\
b : B \qquad d(x) : C\,(i(x)) \quad e(y) : C\,(j(y)) \\
\hline
D\,(j, (a), (x)\ d(x), (y)e(y)) = e(b) : C\,(j(b))
\end{array}
$$

A.1.2   AI.2. Hypotheticals

A.1.2.1   AI.2.a. Hypotheticals and Dependent Objects

The CTT language has also *hypothetical* judgements of the form

$$
B\ type\ (x : A)
$$

Where A is a type which does not depend on any assumptions and B is a type when $x : A$ (the *hypothesis* for B). In the case of sets we have that b is an element of the set B, under the assumption that x is an element of the A:

$$
b : B\ (x : A)\ (\text{more precisely} : b : el(B)\ (x : el\ (A)))
$$

The explicit introduction of hypotheticals carries with it the explicit introduction of appropriate substitution rules. Indeed, if in the example above, $a : A$, then the substitution of x by a in b yields an element of B; and *if $a = c : A$*, then the substitutions of x by a and by c in b are equal elements in B (cfr. Granström [9, pp. 111–112]):

$$
\begin{array}{cc}
\dfrac{a : A \quad b : B\,(x : A)}{b\,(a/x) : B} & \qquad \dfrac{a = c : A \quad b : B\,(x : A)}{b\,(a/x) = b\,(c/x) : B}
\end{array}
$$

As pointed out by Granström [9, p. 112] the form of assertion $b : B\ (x : A)\ (b : el\ (B)\ (x : el\ (A)))$ can be generalized in three directions:

1. Any number of assumptions will be allowed, not just one;
2. The set over which a variable ranges may depend on previously introduced variables;
3. The set B may depend on all introduced variables

Such a list of assumptions will be called a *context*. Thus we might need the forms of assertion

$$b : B \ (\Gamma) \text{ –where } \Gamma \text{ is a context (i.e., a list of assumptions)}$$
$$\Gamma : context$$

In general, a hypothetical judgment has the form

$$x_1 : A_1, x_2 : A_2, \ldots x_n : A_n$$

where we already know that *A_1 is a type, A_2 is a type in the context $x_1 : A_1$, ..., and A_n is a type in the context $x_1 : A_1, x_2 : A_2, \ldots x_{n-1} : A_{n-1}$*:

$$A_1 \ type \quad [\text{depending on no assumption}]$$
$$A_2 \ type \quad (x_1 : A_1)$$
$$\ldots$$
$$A_n \ type \quad (x_1 : A_1, x_2 : A_2, \ldots x_{n-1} : A_{n-1})$$
$$A \ type \quad (x_1 : A_1, x_2 : A_2, \ldots x_n : A_n)$$
$$- - - - - - - - - - - - - - - - - - - -$$
$$x : A \quad (x_1 : A_1, x_2 : A_2, \ldots x_n : A_n, x : A)$$

The rules for substitution and equality are generalized accordingly:
Hypothetical judgements introduce functions from A to B:

$$f(x) : B \ (x : A)$$

It can be read in several ways, for example:

f(x) : B for arbitrary x : A
f(x) : B under the hypothesis x : A
f(x) : B provided x : A
f(x) : B given x : A
f(x) : B if x : A
f(x) : B in the context x : A

It is crucial to notice that the notion of function is intentional rather than an extensional. Indeed, the meaning of an hypothetical function that introduces a function is that whatever element a is substituted for x in ($f(x)$, an element $f(a)$ of B results . Moreover, the equality of two functions defined by establishing that substitutions of equal elements of A result in equal elements of B as regulated by the rules of substitution given above – where $b(x)$ is interpreted as function from A to B.[1]:

---

[1]Cf. Ranta [31, p. 21], Nordström, Petersson and Smith [39, chapter 3.3], Primiero [34, p. 47–55] and Granström [9, p. 77–102].

In addition to domains of individuals, an interpreted scientific language requires propositions. They are introduced in CTT by laying down what counts as proof of a proposition. Accordingly, a proposition is true if there is such a proof. We write

$$A : prop$$

to formalize the judgement that A is a proposition. Propositional functions are introduced by hypothetical judgements. The hypothetical judgement required to introduce propositional functions is of the form:

$$B(x) : prop\ (x : A)$$

that reads, *B(x) is of the type proposition, provided it is applied to elements of the (type-)set A*. The rule by which we produce propositions from propositional functions is the following:

$$\frac{a\ :\ A\quad B(x)\ :\ prop\ (x\ :\ A)}{Ba\ :\ prop}$$

And it requires also of the formulation of an appropriate rule that defines the equivalence relation within the type prop:

$$\frac{a = b\ :\ A\quad B(x)\ :\ prop\ (x\ :\ A)}{Ba = Bb\ :\ prop}$$

The notion of propositional function as hypothetical judgement allows the (intensional) introduction of subsets by separation:

$$\frac{A\ :\ set\quad B(x)\ :\ prop\ (x\ :\ A)\quad b\ :\ A\qquad Bb\ true}{\{x\ :\ A/B(x)\}\ :\ set \qquad\qquad b\ :\ \{x\ :\ A/B(x)\}}$$

This explanation of subsets also justifies the following rules:

$$\frac{b\ :\ \{x\ :\ A/B(x)\}}{b\ :\ A}\qquad\qquad\frac{b\ :\ \{x\ :\ A/B(x)\}}{Bb\ true}$$

What *is given in a context*, the *given contextually-dependent knowledge*, is whatever can be derived from the hypotheses constituting the context. Actually, it is usually distinguished between what is *actually given* in the context (*actual knowledge*), namely the variables themselves and the judgements involving these variables and what is potentially given (*potential knowledge*), namely what can be derived by the rules of type theory from what is actually given. Now, actual and potential knowledge can be increased by extending a given context in ways to be described below.

A.1.2.2 AI.2.b. Extending Hypotheticals

Let us consider once more the hypothetical

$$B(x) : prop\, (x : A)$$

Then, we can produce an *extension of the context by interpretation* by means of definitional equalities such *as a=x : A* yielding

$$B(a) : prop\ (a = x : A)$$

Ranta [31, pp. 135–137] applies it to the study of a literary text where the text is seen as defining a context. That is, as a series of hypothetical judgements that can be interpreted by equating the variables with actual objects:

> An interpretation of Hemingway's short story 'The Battler' might start with the definition
> Nick Adams = Ernest Hemingway : man
> and go on assigning events from the young Hemingway's life to the variable proofs of even propositions asserted in the story. [40, p. 136]

More generally one could extend a context by another context that interprets the variables of the original context in terms of the new ones (cfr. [31, pp. 145–147]). Extensions can in principle induce the growing of knowledge (cfr. Primiero [34, pp. 155–163]). In fact, a context can be enlarged by:

(a) Addition of hypotheses. For instance the context
$\Gamma = (x_1 : A_1, \ldots x_n : A_n)$ is extended to the context
$\Delta = (x_1 : A_1, \ldots x_n : A_n, x_{n+1} : A_{n+1})$. It is clear that everything that is given in Γ is given in the new context as well and thus in the new context we know what we knew in the original one. It may also happen that in the new context proofs are now available that were not at all available in Γ – not even potentially. In this case an increasing grow of knowledge obtains.

(b) Addition of definitions that interpret one of its variables. This the case already mentioned at the start of the paragraph. A more general formulation is the following: the context
$\Gamma = (x_1 : A_1, \ldots x_n : A_n)$ is extended to the context
$\Delta = (\Gamma, x_k = a : A_k)$
So that in the new context every occurrence of x_k is substituted by a. The new context is obtained from Γ by removing the hypothesis $x_k : A_k$ by $a(x_1 \ldots x_n)$. Thus the new context is shorter than the original. Still, this operation furnishes

not only the knowledge of the original context but the value of the one variable reduces the uncertainty within the context.[2]

(c) Addition of a sequence of definitions of all variables in terms of the variables of the new context (the new context need not look the same as the original one). The context

$\Gamma = x_1 : A_1, \ldots x_n : A_n (x_1 \ldots x_{n-1})$ is extended to the context

$\Delta = y_1 : B_1, \ldots y_m : B_m (y_1 \ldots y_{m-1})$ by a mapping $\mathbf{f}$ from Δ to Γ constituted by a sequence of functions such that

$x_1 = f(y_1) \ldots (y_m) : A_1 (\Delta)$

$\ldots$

$x_n = f(y_1) \ldots (y_m) : A_n (f(y_1) \ldots (y_m) \ldots f_{n-1}(y_1) \ldots (y_m)) (\Delta)$

The third operation of extension can be seen as a generalization of the other two (if the new context results by addition of hypotheses we have the first case; if the new context results from the introduction of only one definition, then we have the second case) by *translating* the old context into the new. Thus, the existence of a mapping $\mathbf{f} : \Delta \rightarrow \Gamma$ is usually taken to be the definition of what it is for a context to be an extension of another context.

It might be even argued as Primiero [34, p. 187] does, that this extension amounts to knowledge enlargement in the sense that the new context can show some properties holding in the old context in such a way that new concepts might elucidate the older ones.

A.2 Appendix II: Dialogical Logic and Constructive Type Theory

The dialogical approach to logic is not a specific logical system but rather a rule-based semantic framework in which different logics can be developed, combined and compared.[3] An important point is that the rules that fix meaning are of more than one kind.[4] This feature of its underlying semantics quite often motivated the

---

[2]Ranta [31, p. 146] points out that in a series of lectures Per Martin-Löf showed how the growth of knowledge in experiments can be understood in this way: an unknown quantity is assigned a value, which may depend on other unknown quantities.

[3]For a more thorough presentation, see [41, 42].

[4]The main original papers are collected in [43]. For an historical overview see Lorenz [44]. For a presentation about the initial role of the framework as a foundation for intuitionistic logic, see Felscher [45]. Other papers have been collected more recently in Lorenz [46–48]. A detailed account of recent developments since Rahman [49], can be found in Rahman and Keiff [50], Keiff [51] and Rahman [52]. For the underlying metalogic see Clerbout [53, 54]. For textbook presentations: Kamlah and Lorenzen [55, 56], Lorenzen and Schwemmer [57], Redmond and Fontaine [58] and Rückert [40]. For the key role of dialogic in regaining the link between dialectics and logic, see Rahman and Keiff [59]. Keiff [60–62] and Rahman [63] study Modal Dialogical Logic. Fiutek et al. [64] study the dialogical approach to belief revision. Clerbout, Gorisse and

dialogical approach to be understood as a *pragmatist* semantics. More precisely, in a dialogue two parties argue about a thesis respecting certain fixed rules. The player that states the thesis is called Proponent (**P**), his rival, who contests the thesis is called Opponent (**O**). In its original form, dialogues were designed in such a way that each of the plays end after a finite number of moves with one player winning, while the other loses. Actions or moves in a dialogue are often understood as speech-acts involving *declarative utterances or posits and interrogative utterances or requests*. The point is that the rules of the dialogue do not operate on expressions or sentences isolated from the act of uttering them. The rules are divided into particle rules or rules for logical constants (*Partikelregeln*) and structural rules (*Rahmenregeln*). The structural rules determine the general course of a dialogue game, whereas the particle rules regulate those moves (or utterances) that are requests (to the moves of a rival) and those moves that are answers (to the requests).

Crucial for the dialogical approach are the following points (cfr. Rahman [52]):

1. The distinction between *local* (rules for logical constants) and *global* meaning (included in the structural rules that determine how to play).
2. The player independence of local meaning.
3. The distinction between the play level (local winning or winning of a play) and the strategic level (existence of a winning strategy).
4. A notion of validity that amounts to winning strategy *independently of any model* instead of winning strategy for *every* model.
5. The distinction between non formal and formal plays – the latter notion concerns plays that are played independently of knowing the meaning of the elementary sentences involved in the main thesis.

In the framework of constructive type theory propositions are sets whose elements are called proof-objects. When such a set is not empty, it can be concluded that the proposition has a proof and that it is true. In his 1988 paper [65], Ranta proposed a way to make use of this approach in relation to game-theoretical approaches. Ranta took Hintikka's Game Theoretical Semantics as a case study, but the point does not depend on this particular framework. Ranta's idea was that in the context of game-based approaches, a proposition is a set of winning strategies for the player positing the proposition.[5] Now in game-based approaches, the notion of truth is to be found at the level of such winning strategies. This idea of Ranta's should therefore enable us to apply safely and directly methods taken from constructive type theory to cases of game-based approaches.

Rahman [66] studied Jain Logic in the dialogical framework. Popek [67] develops a dialogical reconstruction of medieval *obligationes*. Rahman and Tulenheimo [68] study the links between GTS and Dialogical Logic. For other books see Redmond [69] – on fiction and dialogic – Fontaine [70] – on intentionality, fiction and dialogues – and Magnier [1] – on dynamic epistemic logic (van Ditmarsch et al. [71]) and legal reasoning in a dialogical framework.

[5]That player can be called Player 1, Myself or Proponent.

But from the perspective of game theoretical approaches, reducing a game to a set of winning strategies is quite unsatisfactory, all the more when it comes to a theory of meaning. This is particularly clear in the dialogical approach in which different levels of meaning are carefully distinguished. There is thus the level of strategies which is a level of meaning analysis, but there is also a level prior to it which is usually called the level of plays. The role of the latter level for developing an analysis is, according to the dialogical approach, crucial, as pointed out by Kuno Lorenz in his 2001 paper [44]:

> [...] for an entity [A] to be a proposition there must exist a dialogue game associated with this entity [...] such that an individual play where A occupies the initial position [...] reaches a final position with either win or loss after a finite number of moves [...]

For this reason we would rather have propositions interpreted as sets of what we shall call *play-objects*, reading an expression

$$p : \varphi$$

as "p is a play-object for φ ".

Thus, Ranta's work on proof objects and strategies constitutes the end not the start of the dialogical project.

A.2.1   The Formation of Propositions

Before delving into the details about play-objects, let us first discuss the issue of the formation of expressions and in particular of propositions in the context of dialogical logic.

In standard dialogical systems, there is a presupposition that the players use well-formed formulas. One can check the well formation at will, but only with the usual meta reasoning by which one checks that the formula indeed observes the definition of wff. The first enrichment we want to make is to allow players to question the status of expressions, in particular to question the status of something as actually standing for a proposition. Thus, we start with rules giving a dialogical explanation of the *formation* of propositions. These are local rules added to the particle rules which give the local meaning of logical constants (see next section).

Let us make a remark before displaying the formation rules. Because the dialogical theory of meaning is based on argumentative interaction, dialogues feature expressions which are not posits of sentences. They also feature requests used for challenges, as illustrated by the formation rules below and the particle rules in the next section. Now, by the *no entity without type* principle, the type of these actions, which we may write "*formation-request*", should be specified during a dialogue. Nevertheless we shall consider that the force symbol $?_F$ already makes the type explicit. Indeed a request in a dialogue should not be confused with a move

Table 7.2 Formation rules

Posit	Challenge [when different challenges are possible, the challenger chooses]	Defence
	$\mathbf{Y}\ ?_{\mathrm{can}}\,\Gamma$	$\mathbf{X}\,!\,a_1 : \Gamma, \mathbf{X}\,!\,a_2 : \Gamma, \ldots$
	Or	$\mathbf{X}$ gives the canonical elements of Γ
$\mathbf{X}\,!\,\Gamma$: set	$\mathbf{Y}\ ?_{\mathrm{gen}}\,\Gamma$	$\mathbf{X}!a_i : \Gamma \Rightarrow a_j : \Gamma$
	Or	$\mathbf{X}$ provides a generation method
	$\mathbf{Y}\ ?_{\mathrm{eq}}\,\Gamma$	$\mathbf{X}$ gives the equality rule for Γ
	$\mathbf{Y}\ ?_{\mathrm{F}\vee 1}$	$\mathbf{X}\,!\,\varphi$: prop
$\mathbf{X}!\varphi \vee \psi$: prop	Or	Respectively
	$\mathbf{Y}\ ?_{\mathrm{F}\vee 2}$	$\mathbf{X}\,!\,\psi$: prop
	$\mathbf{Y}\ ?_{\mathrm{F}\wedge 1}$	$\mathbf{X}\,!\,\varphi$: prop
$\mathbf{X}!\varphi \wedge \psi$: prop	or	Respectively
	$\mathbf{Y}\ ?_{\mathrm{F}\wedge 2}$	$\mathbf{X}\,!\,\psi$: prop
	$\mathbf{Y}\ ?_{\mathrm{F}\rightarrow 1}$	$\mathbf{X}\,!\,\varphi$: prop
$\mathbf{X}!\varphi \rightarrow \psi$: prop	or	Respectively
	$\mathbf{Y}\ ?_{\mathrm{F}\rightarrow 2}$	$\mathbf{X}\,!\,\psi$: prop
	$\mathbf{Y}\ ?_{\mathrm{F}\forall 1}$	$\mathbf{X}\,!\,A$: set
$\mathbf{X}!\,(\forall x : A)\,\varphi(x)$: prop	or	Respectively
	$\mathbf{Y}\ ?_{\mathrm{F}\forall 2}$	$\mathbf{X}\,!\,\varphi(x)$: prop $(x : A)$
	$\mathbf{Y}\ ?_{\mathrm{F}\exists 1}$	$\mathbf{X}\,!\,A$: set
$\mathbf{X}!\,(\exists x : A)\,\varphi(x)$: prop	or	Respectively
	$\mathbf{Y}\ ?_{\mathrm{F}\exists 2}$	$\mathbf{X}\,!\,\varphi(x)$: prop $(x : A)$
$\mathbf{X}\,!\,\mathrm{B}(k)$: prop (for atomic B)	$\mathbf{Y}\ ?_{\mathrm{F}}$	$\mathbf{X}$ sic (n) ($\mathbf{X}$ indicates that $\mathbf{Y}$ posited it in move n)
$\mathbf{X}\,!\,\bot$: prop	–	–

by the means of which it is posited that some entity is of the type request.[6] Hence the way requests are written in rules and dialogues in this work (see Table 7.2).

By definition the falsum symbol $\bot$ is of type prop. A posit $\bot$ cannot therefore be challenged.

The next rule is not formation rules per se but rather a substitution rule.[7] When φ is an elementary sentence, the substitution rule helps explaining the formation of such sentences.

Posit-Substitution

There are two cases in which $\mathbf{Y}$ can ask $\mathbf{X}$ to make a substitution in the context $x_i :$ A_i. The first one is when in a standard play a (list of) variable occurs in a posit with a proviso. Then the challenger posits an instantiation of the proviso (see Table 7.3).

---

[6]Such a move could be written as $?_{\mathrm{F}\vee 1}$: *formation-request*.

[7]It is an application of the original rule from CTT given in Ranta [31, p. 30].

Table 7.3 Posit-substitution I

Posit	Challenge	Defence
$\mathbf{X}\,!\,\pi(x_1, \ldots x_n)\ (x_i : A_i)$	$\mathbf{Y}\,!\,t_1 : A_1 \ldots t_n : A_n$	$\mathbf{X}\,!\,\pi(\tau_1, \ldots, \tau_n)$

Table 7.4 Posit-substitution II

Posit	Challenge	Defence
$\mathbf{X}\,!\,\pi(\tau_1, \ldots \tau_n)\ (\tau_i : A_i)$	$\mathbf{Y}\,!\,\tau_1 : A_1 \ldots \tau_n : A_n$	$\mathbf{X}\,!\,\pi(\tau_1, \ldots, \tau_n)$

The second case is in a formation-play. In such a play the challenger simply posits the whole assumption as in move 7 of Table 7.4.

Remarks on the Formation Dialogues

(a) *Conditional formation posits:*

One crucial feature of the formation rules is that they allow displaying the syntactic and semantic presuppositions of a given thesis and thus can be examined by the Opponent before the actual dialogue on the thesis is run. Thus if the thesis amounts to positing, say, φ, then before an attack is launched, the opponent can asked for its formation. The defence of the formation of φ might conduce the Proponent to posit that φ is a proposition, under the condition that it is conceded that, say A is a set. In such a situation the Opponent might accept to concede A is a set, but only after P has displayed the constitution of A.

(b) *Elementary sentences, definitional consistency and material-analytic dialogues:*

If we follow thoroughly the idea of formation rules, one should allow elementary sentences to be challenged: by the formation rules. The defence will make use of the applications of adequate conceded predicator rules (if there are any such concessions). Therefore, what will happen is that the challenge on elementary sentence is based on the definitional consistency in use of the conceded the predicator rules. This is what we think material-dialogues are about: definitional consistency dialogues. This leads to the following material analytic rule *for formation dialogues:*

> **O**'s elementary sentences cannot be challenged, however **O** can challenge an elementary sentence (posited by **P**) iff herself (the opponent) did not posit it before.

Remark Once the proponent forced the Opponent to concede the elementary sentence in the formation dialogue, the dialogue will proceed making use of the copy-cat strategy.

(c) *Indoor-* versus *outdoor-games:*

Hintikka [72, pp. 77–82], who acknowledges the close links between dialogical logic and GTS launched an attack against the philosophical foundations of dialogic because of their *indoor-* or purely formal approach to meaning as use. He argues that *formal proof* games are not of very much help in accomplishing the task of linking the linguistic rules of meaning with the real world.

> In contrast to our games of seeking and finding, the games of Lorenzen and Stegmüller are 'dialogical games' which are played 'indoors' by means of verbal 'challenges' and 'responses'. [. . .].
>
> [. . .]. If one is merely interested in suitable technical problems in logic, there may not be much to choose between the two types of games. However, from a philosophical point of view, the difference seems to be absolutely crucial. Only considerations which pertain to 'games of exploring the world' can be hoped to throw any light on the role of our logical concepts in the meaningful use of language. [72, p. 81].

Rahman and Keiff [50, p. 379] pointed out that *formal proof*, that is validity, does not in the dialogical frame provide meaning either: it is rather the other way round, i.e. formal plays furnish the basis for the notion of dialogical validity (that amounts to the notion of a *winning **P**-strategy*).

By way of illustration, we present a dialogue where the Proponent posits the thesis $(\forall x : A)\, B(x) \rightarrow C(x)$: prop given that A : set, $B(x)$: prop $(x : A)$ and $C(x)$: prop $(x : A)$, where the three provisos appear as initial concessions by the Opponent.[8] Good form demands that we first present the structural rules which define the conditions under which a play can start, proceed and end. But we leave them for the next section. They are not necessary to understand the dialogue in Table 7.5.

Explanations **Move I to III: O** concedes that A is a set and that $B(x)$ and $C(x)$ are propositions provided x is an element of A,

Explanations **Move 0**: **P** posits that the main sentence, universally quantified, is a proposition (under the concessions made by **O**),

Explanations **Moves 1 and 2**: the players choose their repetition ranks,

Explanations **Move 3**: **O** challenges the thesis a first time by asking the left-hand part as specified by the formation rule for universal quantification,

Table 7.5 A Formation-play

	O			P	
I	$!\,A$: set				
II	$!\,B(x)$: prop $(x : A)$				
III	$!\,C(x)$: prop $(x : A)$				
				$!\ (\forall x : A)\, B(x) \rightarrow C(x)$: prop	0
1	n := 2			m := 2	2
3	$?_{\mathrm{F}\forall 1}$	(0)		$!\,A$: set	4
5	$?_{\mathrm{F}\forall 2}$	(0)		$!\,B(x) \rightarrow C(x)$: prop $(x : A)$	6
7	$!\,x : A$	(6)		$!\,B(x) \rightarrow C(x)$: prop	8
9	$?_{\mathrm{F}\rightarrow 1}$	(8)		$B(x)$: prop	12
11	$!\,B(x)$: prop		(II)	$!\,x : A$	10
13	$?_{\mathrm{F}\rightarrow 2}$	(8)		$!\,C(x)$: prop	16
15	$!\,C(x)$: prop		(III)	$!\,x : A$	14

---

[8]The example comes from Ranta [31, p. 31].

Explanations **Move 4**: **P** responds by positing that A is a set. This has already been
   granted with the premise I so **P** can make this move while respecting the
   Formal rule,

Explanations **Move 5**: **O** challenges the thesis again, this time asking for the right-
   hand part,[9]

Explanations **Move 6**: **P** responds, positing that $B(x) \rightarrow C(x)$ is a proposition
   provided $x : A$,

Explanations **Move 7**: **O** uses the substitution rule to challenge move 6 by granting
   the proviso,

Explanations **Move 8**: **P** responds by positing that $B(x) \rightarrow C(x)$ is a proposition,

Explanations **Move 9**: **O** then challenges move 8 a first time by asking the left-hand
   part as specified by the formation rule for material implication.

In order to defend **P** needs to make an elementary move. But since **O** has not
played it yet, **P** cannot defend at this point. Thus:

Move 10: **P** launches a counterattack against assumption II by applying the first
   case of the substitution rule,

Move 11: **O** answers move 10 and posits that $B(x)$ is a proposition,

Move 12: **P** can now defend in reaction to move 9,

Move 13: **O** challenges move 8 a second time, this time requiring the right-hand
   part of the conditional,

Move 14: **P** launches a counterattack and challenges assumption III by applying
   again the first case of the substitution rule

Move 15: **O** defends by positing that $C(x)$ is a proposition,

Move 16: **P** can now answer to the request of move 13 and win the dialogue (**O** has
   no further move).

From the view point of building a winning strategy, the Proponent's victory only
shows that the thesis is justified *in this particular play*. To build a winning strategy
we must run all the relevant plays for this thesis under these concessions.

Now that the dialogical account of formation rules has been clarified, we may
develop further our analysis of plays by introducing play-objects.

A.2.2   Play Objects

The idea is now to design dialogical games in which the players' posits are of the
form "$p : \varphi$" and acquire their meaning in the way they are used in the game – i.e.,
how they are challenged and defended. This requires, among others, to analyse the
form of a given play-object p, which depends on φ, and how a play-object can be
obtained from other, simpler, play-objects. The standard dialogical semantics for

---

[9]This can be done because **O** has chosen 2 for her repetition rank.

logical constants gives us the needed information for this purpose. The main logical constant of the expression at stake provides the basic information as to what a play-object for that expression consists of:

A play for $X!\varphi \vee \psi$ is obtained from two plays p_1 and p_2, where p_1 is a play for $X \: ! \: \varphi$ and p_2 is a play for $X \: ! \: \psi$. According to the particle rule for disjunction, it is the player X who can switch from p_1 to p_2 and vice-versa.

A play for $X!\varphi \wedge \psi$ is obtained similarly, except that it is the player Y who can switch from p_1 to p_2.

A play for $X!\varphi \rightarrow \psi$ is obtained from two plays p_1 and p_2, where p_1 is a play for $Y \: ! \: \varphi$ and p_2 is a play for $X \: ! \: \psi$. It is the player X who can switch from p_1 to p_2.

The standard dialogical particle rule for negation rests on the interpretation of $\neg\varphi$ as an abbreviation for $\varphi \rightarrow \bot$, although it is usually left implicit. It follows that a play for $X!\neg\varphi$ is also of the form of a conditional, where p_1 is a play for $Y \: ! \: \varphi$ and p_2 is a play for $X! \bot$, and where X can switch from p_1 to p_2. Notice that this approach covers the standard game-theoretical interpretation of negation as role-switch: p_1 is a play for a Y move.

As for quantifiers, we give a detailed discussion after the particle rules further on. For now, we would like to point out that, just like what is done in constructive type theory, we are dealing with quantifiers for which the type of the bound variable is always specified. We thus consider expressions of the form $(Qx : A) \: \varphi$, where Q is a quantifier symbol (see Table 7.6).

Notice that we have added for each logical constant a challenge of the form 'Y $?_{prop}$' by which the challenger questions the fact that the expression at the right-hand side of the semi-colon is a proposition. This makes the connection with the formation rules given in Table 7.2 via X's defence. More details are given in the discussion after the structural rules.

It may happen that the form of a play-object is not explicit at first. In such cases we deal with expressions of the form, e.g., "$p \: : \: \varphi \wedge \psi$". We may then use expressions of the form $L^{\wedge}(p)$ and $R^{\wedge}(p)$ – which we call *instructions* – in the relevant defences. Their respective interpretations are "take the left part of p" and "take the right part of p". In instructions we indicate the logical constant at stake. First it keeps the formulations explicit enough, in particular in the case of embedded instructions. More importantly we must keep in mind that there are important differences between play-objects depending on the logical constant. Remember for example that in the case of conjunction the play-object is a pair, but it is not in the case of disjunction. Thus $L^{\wedge}(p)$ and $L^{\vee}(p)$, say, are actually different things and the notation takes that into account. More precisely,

given a play object p for the disjunction – composed by two play objects such that each of them constitutes a *sufficient* play object for the disjunction – the expression $L^{\vee}(p)$ ($R^{\vee}(p)$) instructs the *defender* to choose the play object for the left (right) side of the disjunction.

given a play the object p for the conjunction – composed by two play objects such that in order to constitute a play object for the conjunction each of them is *necessary* – the expression $L^{\wedge}(p)$ ($R^{\wedge}(p)$) instructs the *challenger* to choose the play object for the left (right) side of the conjunction.

Let us focus on the rules for quantifiers. Dialogical semantics highlights the fact that there are two distinct moments when considering the meaning of quantifiers: the choice of a suitable substitution term for the bound variable, and the instantiation

Table 7.6 Local rules

Posit	Challenge	Defence
$X\,!\,\varphi$ (where no play-object has been specified for φ)	$Y\,?\,play\text{-}object$	$X\,!\,p:\varphi$
	$Y\,?_{prop}$	$X!\varphi \vee \psi : \text{prop}$
$X!\,p:\varphi \vee \psi$	$Y\,?\,[\varphi, \psi]$	$X!\,L^{\vee}(p):\varphi$ Or $X!\,R^{\vee}(p):\psi$ **[the defender has the choice]**
$X!\,p:\varphi \wedge \psi$	$Y\,?_{prop}$	$X!\varphi \wedge \psi : \text{propb}$
	$Y\,?[\varphi]$	$X!\,L^{\wedge}(p):\varphi$
	or	respectively
	$Y\,?[\psi]$ **[the challenger has the choice]**	$X!\,R^{\wedge}(p):\psi$
$X!\,p:\varphi \rightarrow \psi$	$Y\,?_{prop}$	$X!\varphi \rightarrow \psi : \text{prop}$
	$Y!\,L^{\rightarrow}(p):\varphi$	$X!\,R^{\rightarrow}(p):\psi$
$X!\,p:\neg\varphi$	$Y\,?_{prop}$	$X!\neg\varphi : \text{prop}$
	$Y!\,L^{\perp}(p):\varphi$	$X!\,R^{\perp}(p):\perp$
$X!\,p:(\exists x:A)\,\varphi$	$Y\,?_{prop}$	$X!\,(\exists x:A)\,\varphi : \text{prop}$
	$Y\,?_{L}$	$X!\,L^{\exists}(p):A$
	or	respectively
	$Y\,?_{R}$ **[the challenger has the choice]**	$X!\,R^{\exists}(p):\varphi\,(L(p))$
$X!\,p:\left\{x:A\,\middle|\,\varphi\right\}$	$Y\,?_{L}$	$X\,!\,L^{\{\ldots\}}(p):A$
	or	respectively
	$Y\,?_{R}$ **[the challenger has the choice]**	$X\,!\,R^{\{\ldots\}}(p):\varphi(L(p))$
$X!\,p:(\forall x:A)\,\varphi$	$Y\,?_{prop}$	$X!\,(\forall x:A)\,\varphi : \text{prop}$
	$Y\,L^{\forall}(p):A$	$X!\,R^{\forall}(p):\varphi\,(L(p))$
$X\,!\,p:B(k)$ (for atomic B)	$Y\,?_{prop}$	$X\,!\,B(k) : \text{prop}$
	$Y\,?$	$X\,!\,\text{sic}\,(n)$ (X indicates that Y posited it at move n)

of the formula after replacing the bound variable with the chosen substitution. But at the same time in the standard dialogical approach there is some sort of presupposition that every quantifier symbol ranges on a unique kind of objects. Now, things are different in the context of the explicit language we borrow from CTT. Quantification is always relative to a set, and there are sets of many different kinds of objects (for example: sets of individuals, sets of pairs, sets of functions, etc.).

Thanks to the instructions we can give a general form for the particle rules. It is in a third, later, moment that the kind of object is specified, when instructions are "resolved" by means of the structural rule SR4.1 below.

Constructive type theory makes it clear that as soon as propositions are thought of as sets, there is a basic similarity between conjunction and existential quantifier on the one hand and material implication and universal quantifier on the other hand. Briefly, the point is that they are formed in similar ways and their elements are generated by the same kind of operations.[10] In our approach, this similarity manifests itself in the fact that a play-object for an existentially quantified expression is of the same form as a play-object for a conjunction. Similarly, a play-object for a universally quantified expression is of the same form as one for a material implication.[11]

The particle rule just before the one for universal quantification is a novelty in the dialogical approach. It involves expressions commonly used in constructive type theory to deal with separated subsets. The idea is to understand *those elements of A such that* φ as expressing that at least one element $L^{\{\cdots\}}(p)$ of A witnesses $\varphi(L^{\{\cdots\}}(p))$. The same correspondence that linked conjunction and existential quantification now appears.[12] This is not surprising since such posits actually have an existential aspect: in $\{x : A \mid \varphi\}$ the left part "$x : A$" signals the existence of a play-object. Let us point out that since the expression stands for a set there is no presupposition that it is a proposition when **X** makes the posit. This is why it cannot be challenged with the request "$?_{\text{prop}}$".

A.2.3 The Development of a Play

In this section we deal with the so-called *called structural rules*. These rules govern the way plays globally proceed and are therefore an important aspect of a dialogical semantics. We work with the following structural rules:

---

[10]More precisely, conjunction and existential quantifier are two particular cases of the Σ operator (disjoint union of sets), whereas material implication and universal quantifier are two particular cases of the Π operator (indexed product on sets). See for example Ranta [31, chapter 2].

[11]Still, if we are playing with classical structural rules, there is a slight difference between material implication and universal quantification which we take from Ranta [31, Table 2.3], namely that in the second case p_2 always depends on p_1.

[12]As pointed out in Martin-Löf [38], subset separation is another case of the Σ operator. See in particular p. 53:

"Let A be a set and B(x) a proposition for x ε A. We want to define the set of all a ε A such that B(a) holds (which is usually written $\{x \varepsilon A : B(x)\}$). To have an element a ε A such that B(a) holds means to have an element a ε A together with a proof of B(a), namely an element b ε B(a). So the elements of the set of all elements of A satisfying B(x) are pairs (a; b) with b ε B(a), i.e. elements of (Σx ε A)B(x). Then the Σ-rules play the role of the comprehension axiom (or the separation principle in ZF)."

SR0 (Starting rule) Any dialogue starts with the Proponent positing the thesis. After that the players each choose a positive integer called repetition ranks.

SR1i (Intuitionistic Development rule) Players move alternately. After the repetition ranks have been chosen, each move is a challenge or a defence in reaction to a previous move, in accordance with the particle rules. The repetition rank of a player bounds the number of challenges he can play in reaction to a same move. Players can answer only against *the last non-answered* challenge by the adversary.

[**SR1c (Classical Development rule)** Players move alternately. After the repetition ranks have been chosen, each move is a challenge or a defence in reaction to a previous move, in accordance with the particle rules. The repetition rank of a player bounds the number of challenges and defences he can play in reaction to a same move.]

SR2 (Formation first) O starts by challenging the thesis with the request '?$_{\text{prop}}$'. The game then proceeds by applying the formation rules first in order to check that the thesis is indeed a proposition. After that the Opponent is free to use the other local rules insofar as the other structural rules allow it.

SR3 (Modified Formal rule) O's elementary sentences cannot be challenged, however O can challenge an elementary sentence (posited by P) iff herself (the opponent) did not posit it before.

SR4.1 (Resolution of instructions) Whenever a player posits a move where instructions I_1, ..., I_n occur, the other player can ask him to replace these instructions (or some of them) by suitable play-objects.

If the instruction (or list of instructions) occurs at the right of the colon and the posit is the tail of an universally quantified sentence or of an implication (so that these instructions occur at the left of the colon in the posit of the head of the implication), then it is the challenger who can choose the play-object – in these cases the player who challenges the instruction is also the challenger of the universal quantifier and/or of the implication.

Otherwise it is the defender of the instructions who chooses the suitable play-object (see Table 7.7).

Important Remark In the case of embedded instructions $I_1(\dots(I_k)\dots)$, the substitutions are thought as being carried out from I_k to I_1: first substitute I_k with some

Table 7.7 Instructions I

Posit	Challenge	Defence
$X\,!\,\pi(I_1,\dots,I_n)$	$Y\,I_1,\dots,I_m =?\ (m \le n)$	$X\,!\,\pi(b_1,\dots,b_m)$
		- if the instruction that occurs at the right of the colon is the tail of either a universal or an implication (such that $I_1,\dots,I_n$ also occur at the left of the colon in the posit of the head),
		then $b_1,\dots,b_m$ **are chosen by the challenger**
		- Otherwise **the defender chooses**

Table 7.8 Instructions II

Posit	Challenge	Defence
Player 1 ! $\pi_i(I)$	**Y**? b/I	**X** ! $\pi_j(b)$
Player 2 $I = ?$		
Player 3 ! $\pi_i(b)$		
...		
X-!- $\pi_j(I)$		

play-object b_k, then $I_{k-1}(b_k)$ with b_{k-1} ... until $I_1(b_2)$. If such a progressive substitution has actually been carried out once, a player can then replace $I_1(\ldots (I_k) \ldots)$ directly.

SR4.2 (Substitution of instructions) When during the play the play-object b has been chosen by any of both players for an instruction I, and player **X** posits $!\pi(I))$, then the antagonist can ask to substitute I with b in any posit **X** ! $\pi(I)$ (see Table 7.8).

The idea is that the resolution of an instruction in a move yields a certain play-object for some substitution term, and therefore the same play-object can be assumed to result for any other occurrence of the same substitution term: instructions are functions after all and as such they must yield the same play-object for the same substitution term.

SR5 (Winning rule for dialogues) For any p, a player who posits "$p : \bot$" looses the current dialogue. Otherwise the player who makes the last move in a dialogue wins it.

In the rules we just gave there are some additions, namely those numbered SR2 and SR4 here, and also the first part of the winning rule. Since we made explicit the use of $\bot$ in our games, we need to add a rule for it: the point is that positing *falsum* leads to immediate loss; we could say that it amounts to a withdrawal (see Keiff [61]).

We need the rules SR4.1 and SR4.2 because of some features of CTT's explicit language. In CTT it is possible to account for questions of dependency, scope, etc directly at the level of the language. In this way various puzzles, such as anaphora, get a convincing and successful treatment. The typical example, which we consider below, is the so-called *donkey sentence* "Every man who owns a donkey beats it". The two rules give a mean to account for the way play-objects can be ascribed to what we have called instructions. See the dialogue below for an application.

The rule SR2 is consistent with the common practice in CTT to start demonstrations by checking or establishing aspects related to the formation of propositions before proving their truth. Notice that this step also covers the formation of sets – membership, generation of elements, etc. – which occur in hypothetical posits and in quantifiers. This is achieved in dialogues by means of rule SR2 which requires that in a dialogue the players first deal with aspects related to formation rules. With this we introduce some resemblance between our games and the CTT approach that makes the task of investigating their connections easier. However, it looks like we could do with a liberalized version of this rule. Because of the number of rules we have introduced, a careful verification of this is a delicate task that we will not carry out in this paper. For now let us simply mention that it looks sensible in the context of dialogues to let the process related to formation rules be more freely combined with the development of a play on the thesis. In fact it does seem perfectly consistent with actual practices of interaction to question the status of expressions once they are introduced in the course of the game. Suppose for example player X has posited '$p : \varphi \vee \psi$'. As soon as he has posited the disjunction to be a proposition – i.e., as soon as he has posited '$\varphi \vee \psi$: prop' – the other player knows how to challenge the disjunction and should be free to keep on exploring the formation of the expression *or* to challenge the first posit. The point is that in a way it seems to make more sense to check whether φ is a proposition or not after (if) X posits it in order to defend the disjunction. Doing so in a 'monological' framework such as CTT would probably bring various confusions, but the dialogical approach to meaning should allow this additional dynamic aspect quite naturally.

A.2.4   *From Play-Objects to Strategies*

We have explained that the view of propositions as sets of winning strategies overlooks the level of plays and that an account more faithful to the dialogical approach to meaning is that of propositions as sets of play-objects. But play-objects are not the dialogical counterparts of CTT proof-objects, and thus are not enough to establish the connection between the dialogical and the CTT approaches.

The local rules of our games – that is, the formation rules together with the particle rules – present some resemblances with the CTT rules, especially if we read the dialogical rules backwards. But in spite of the resemblances, play-objects are in fact very different from CTT proof-objects. The case where the difference is obvious is implication – and thus universal quantification which is similar. In the CTT approach, a proof-object for an implication is a lambda-abstract, and a proof-object of the tail of the implication is obtained by applying the function to the proof-object of the head. But in our account with play-objects, nothing requires that the play-object for the right-hand part is obtained by an application of some function.

From this simple observation it is clear that the connection between our games and CTT is not to be found at the level of plays. In fact it is well-known that the connection between dialogues and proofs is to be found at the level of strategies: see for example Rahman, Clerbout and Keiff [73] for a discussion in relation to natural deduction. Even without the question of the relation with CTT, the task of describing and explaining the level of strategies is due since it is a proper and important level of meaning analysis in the dialogical framework.

A strategy for a player is often defined as a function from the set of non-terminal plays where it is this player's turn to move to the set of possible moves for this player. Equivalently, a strategy can be defined as the set of plays which result when the player follows the strategy. From this we propose to consider strategies as certain sets of play-objects. On the one hand they are different from propositions insofar as a proposition is the set of all possible play-objects for it, whereas any play-object cannot be a member of a given strategy. But on the other hand it is clear that every play-object in a strategy for a proposition A is also in A itself. Thus, a strategy for A is a certain subset of A. This view seems to comply with the dialogical approach according to which the level of strategies is part of the meaning of expressions but does not cover it entirely.

Summing up, we have play-objects which carry the interactive aspects of the meaning-explanations. A proposition is the set of all possible play-objects for it, and a strategy in a game about this proposition is some subset of play-objects for it.

Three important questions must then be addressed. First of all, any subset of A should not count as a strategy for A. So our first question is: what are the conditions that a set of play-objects must observe in order to be called a strategy. Also, the connection between dialogical games and proofs relates to winning strategies for the Proponent. So the second question is: what additional constraints do we need for a strategy to be a winning one? Answering to these questions should lead us to a good understanding of what counts as a canonical (winning) strategy. On this topic, an important remark is that the move from uninterpreted to interpreted languages results in a loss of generality. The clearest illustration is the case of existential quantification. By the particle rule a player making a posit of the form "! p : ($\exists x$: $A)\varphi$" must be ready to provide an element of the set A. If the Proponent is the one making the posit, he needs some previous concession by the Opponent in order to be able to provide an element of A. This means that there cannot be a winning **P**-strategy for posits of this form in the absence of preliminary concessions about the quantification set(s). In other words the dialogue games we have introduced in this paper are in any case not suitable yet to get general validity. To move to validity, an abstraction process must still be worked out such as the one described by Sundholm [29, pp. 33–35]. The dialogical perspective of the abstraction process will presumably involve a more general approach to the copy-cat strategy triggered by the formal rule.

The third question to tackle when moving to the level of strategies is: what are the generation rules for strategies? In other words: what are the operations which can be used to obtain new strategies from already available strategies. In relation to this last question, Ranta [31] proposed to use the same operations that are used in CTT for proof-objects. For example, a winning strategy for A∧B is a pair made of a winning strategy for A and of a winning strategy for B. This certainly makes the connection between winning strategies and proof-objects straightforward. However at first sight it seems a little too simplistic. While it is obvious that (winning) strategies for $A \wedge B$ must be obtained from (wining) strategies for A and for **B**, it seems unsatisfactory to conclude that a strategy for a $A \wedge B$ is a set of pairs of strategies. We would rather keep the idea of the strategy as a set of play-objects. The point would then be that, in the case of $A \wedge B$, the play-objects which are members of the set are obtained from play-objects for A and for B.

Let us finish with a partial answer to the first two questions. We present a procedure by which one can search for (the description of) a winning **P**-strategy in a game.[13] However, as will be clear below, the object(s) which can be obtained by this procedure do not exactly meet the requirements we have listed above. The procedure goes through the construction of expressions similar to the full explicit description of play-objects, but with an important difference: the sequences of moves they represent are not rigorously observing the rules. The reason from this is that on the one hand – for reasons we explain below – we start with the assumption that the Opponent's rank is set to be 1 while on the other hand we allow expansions of the starting expression that **O** should actually not be able to trigger with rank 1. Let us now give some explanations.

First of all, one might wonder why we consider the Opponent's rank to be set beforehand since, strictly speaking, every possible choice of rank for **O** should be considered in a **P**-strategy. Here we rely on the fact that in order to know whether there is a winning **P**-strategy in a given game it is enough to check the case where **O** chooses rank 1 (see Clerbout [53]). Actually other aspects of the procedure, such as the particular choices of individual constants taken for expansions in Steps 2.4 and 2.5, are motivated by considerations taken from the demonstration in Clerbout [53, Chap. 2].

Now, in relation to the second point, it would have been more faithful to the considerations above to explain how alternative ways for the Opponent to play can be built and taken into account, instead of allowing illegal expansions of the starting play. But this is precisely what remains to be done to answer accurately to the three questions we have raised above. The point is that it is a very delicate task to give a procedure which would produce alternatives to the starting play: for this first version we give a flavour of the result we aim at. One of the difficulties we will have to overcome is to keep track of which play-objects have already been counted as belonging to a given set. The procedure below avoids the difficulty by 'merging', so to speak, the various play-objects that would be selected as members of the strategy.

Let us now move to the procedure. As we have explained, the Opponent's rank is 1. As for the Proponent's rank, we assume for now that it is big enough to let **P** keep on playing after an expansion is made: the actual value of his rank can be determined once the procedure ends, when it is possible to count the total number of challenges and defences he made.

Suppose then that we have a play won by **P** in a given game, and that its fully explicit description is given by the play-object ρ.

Preliminaries We say that **O** *makes a decision* in ρ in the following cases:

(I) She challenges a conjunction: she chooses which conjunct to ask for.
(II) She defends a disjunction: she chooses which disjunct to give.

---

[13]The procedure is inspired by the presentation of *strategic games* in Rahman and Keiff [50], Sect. 2.4].

(III) She counter-attacks (or: defends) after a **P**-challenge on a material implication *without defending* (or: *counter-attacking*) *afterwards*.

(IV) She challenges a universal quantifier: she chooses an individual in the set.

(V) She defends an existential quantifier: she chooses an individual in the set.

N.B.: because it is an expression such as the one labelled (B') in the previous Section, ρ actually carries all the information needed to know whether there are such **O**-decisions and where they occur.

Moreover, we say that a move M *depends on* move M' if there is a chain of applications of game (particle) rules from M' to M.

Procedure

1. If there is no (remaining) non-used decision made by **O** in ρ then go to Step 6. Otherwise go to the next step.

2. Take the latest non-used decision d made by **O** in ρ and, depending on the case, apply one of the following and afterwards go to Step 3:

   2.1. If d is a challenge against a conjunction, then expand ρ with the other challenge. That is, take[14] $\rho' = \rho^\cap \mathbf{O}?_L$ (resp. $?_R$) given that $\mathbf{O}?_R$ (resp. $?_L$) occurs in ρ. The game then proceeds as if the first challenge had not taken place: moves depending from the first challenge are forbidden to both players.

   2.2. If f is a defence for a disjunction, then expand ρ with the other disjunct. That is, take $\rho' = \rho^\cap \mathbf{O}!\mathrm{L}^\vee(p) : \varphi$ (resp. $\mathrm{R}^\vee(p) : \psi$) given that $\mathbf{O}!\mathrm{R}^\vee(p) : \psi$ (resp. $\mathrm{L}^\vee(p) : \varphi$) already occurs in ρ. The game then proceeds as if the first defence had not taken place: moves depending from the first defence are forbidden to both players.

   2.3. If d is a counter-attack (resp. a defence) in reaction to a **P**-challenge on a material implication, then expand ρ with the defence (resp. the counter-attack). That is, take $\rho' = \rho^\cap \mathbf{O}!\mathrm{R}^\rightarrow(p) : \psi$ (resp. $\mathbf{O}?\dots$). The game then proceeds as if the counter-attack (resp. the defence) had not taken place: moves depending from it are forbidden to both players.

   2.4. If d is a challenge against a universal quantifier, then we distinguish cases:

      2.4.1. The individual from the set, say $a : A$, chosen at d is new. Then for each other individual a_i in A – if any – occurring previously in ρ, expand ρ with the choice of this individual. That is, take $\rho'_i = \rho^\cap \mathbf{O}!a_i : A$ for each a_i.

      For each such expansion, the game then proceeds as if the first challenge had not taken place: moves depending from it are forbidden to both players.

      2.4.2. The individual chosen at d is *not* new. Then:

---

[14] As should be obvious form the context, '$\cap$' is a concatenation operator.

(a) Expand ρ with a challenge where **O** chooses a new individual. That is, take $\rho'_a = \rho^\cap\mathbf{O}!a : A$ where a is new.

(b) Also, for each other individual a_i of A – if any – occurring previously in ρ, take $\rho'_i = \rho^\cap\mathbf{O}!a_i : A$.

For each such expansion, the game then proceeds as if the first challenge had not taken place: moves depending from it are forbidden to both players.

2.5. If d is a defence of an existential quantifier, then we distinguish cases:

2.5.1. The individual of the set, say $a : A$, chosen at d is new. Then for each other individual a_i in A – if any – occurring previously in ρ, expand ρ with the choice of this individual. That is, take $\rho'_i = \rho^\cap\mathbf{O}!\mathbf{R}^\exists(p) : \varphi(a_i)$ for each a_i.

For each such expansion, the game then proceeds as if the first defence had not taken place: moves depending from it are forbidden to both players.

2.5.2. The individual chosen at d is not new. Then:

(a) Expand ρ with a challenge where **O** chooses a new individual. That is, take $\rho'_a = \rho^\cap\mathbf{O}!\mathbf{R}^\exists(p) : \varphi(a)$ where a is new.

(b) Also, for each other individual a_i of A – if any – occurring previously in ρ, take $\rho'_i = \rho^\cap\mathbf{O}!\mathbf{R}^\exists(p) : \varphi(a_i)$.

For each such expansion, the game then proceeds as if the first defence had not taken place: moves depending from it are forbidden to both players.

3. Name the resulting sequence(s) $\rho*$ (or $\rho*i$ if relevant). Mark d as used and go to the next step.

4. If $\rho*$ (or one of the $\rho*i$) is O-terminal then stop. Take another play won by P and go back to Step 1. Otherwise go to the next step.

5. Take the next non-used **O**-decision in ρ and repeat Step 2 but by expanding $\rho*$ (or each of the $\rho*i$) instead of ρ.

When there are no non-used **O**-decision left, go to Step 6.

6. Call the sequences obtained $\rho^\sigma{}_i$. For each of these take its **O**-permutations, namely the sequences which are the same up to the order of **O**-moves and still observe the game rules.

The set of all the $\rho^\sigma{}_i$ and their **O**-permutations provides a description of a **P**-strategy. If all of these are **P**-terminal then the strategic-object is **P**-winning and there is a winning **P**-strategy in the game at stake.

Important Remark Step 4 makes the procedure a method to search for descriptions of winning **P**-strategies. If one of the expanded play-objects is not won by him, the procedure stops and must be started again with another starting play(−object). Notice that the procedure will keep on searching until a winning **P**-strategy is described. A consequence is that if there is no such strategy in the game the procedure will not accurately determine it and will keep on searching indefinitely. This is consistent with the semi-decidability of first-order dialogical games – and of predicate logic. See Clerbout [53, Chap. 3].

A.2.5 Equality

Sometimes, the following simplified rule for substitution for equalities is useful (see Table 7.9) – for a complete description see following sections – for a thorough study of definitional equality in dialogical logic see [41]

Table 7.9 Simplified equality-rule

Posit	Challenge	Defence
$\mathbf{X}-!-\pi_n\left(\tau_i = \tau_j\right)$	$\mathbf{Y}-?-\tau_f/\tau_i?,$	$\mathbf{X}-!-\pi_m\left[\tau_i\right]$
$\mathbf{X}-!-\pi_m\left[\tau_i\right]$ where '$[\tau_i]$' indicates that τ_i occurs in the judgement π_m $\mathbf{X}$ utters $\pi_n(\tau_i = \tau_j)$ and $\pi_m[\tau_i]$	(the challenger asks X to replace in φ τ_i with τ_i)	(X carries out the required substitution of τ_f for τ_i)

A.2.5.1 Definitional Equality

Table 7.10 Reflexivity within *Set*

Posit	Challenge	Defence
$\mathbf{X}-!-A : set$	$\mathbf{Y}-?_{\text{set}}-refl$	$\mathbf{X}-!-A = A : set$

Table 7.11 Symmetry within *set*

Posit	Challenge	Defence
$\mathbf{X}-!-A = B : set$	$\mathbf{Y}-?_{\text{B}}-symm$	$\mathbf{X}-!-B = A : set$

Table 7.12 Transitivity within *set*

Posit	Challenge	Defence
$\mathbf{X}-!-A = B : set$ $\mathbf{X}-!-B = C : set$	$\mathbf{Y}-?_{\text{A}}-trans$	$\mathbf{X}-!-A = C : set$

Table 7.13 Reflexivity within *A*

Posit	Challenge	Defence
$\mathbf{X}-!-a : A$	$\mathbf{Y}-?\ a\ refl$	$\mathbf{X}-!-a = a : A$

Table 7.14 Symmetry within *A*

Posit	Challenge	Defence
$\mathbf{X}-!-a = b : A$	$\mathbf{Y}-?_b-symm$	$\mathbf{X}-!-b = a : A$

Table 7.15 Transitivity within A

Posit	Challenge	Defence
$\mathbf{X}-!-a=b:A$ $\mathbf{X}-!-b=c:A$	$\mathbf{Y}\text{-}?_a\text{-}\ trans$	$\mathbf{X}-!-a=c:A$

Table 7.16 Simple extensionality

Posit	Challenge	Defence
$\mathbf{X}-!-A=B:set$	$\mathbf{Y}-?_{\text{ext}}-a:A$	$\mathbf{X}-!-a:B$

Table 7.17 Double extensionality

Posit	Challenge	Defence
$\mathbf{X}-!-A=B:set$	$\mathbf{Y}-?_{\text{d}-\text{ext}-}a=b:A$	$\mathbf{X}-!-a=b:B$

Table 7.18 Substitution $B(x)$

Posit	Challenge	Defence
$\mathbf{X}-!-B(x):set\ (x:A)$	$\mathbf{Y}-?_{B(x)-\text{subst}}\ a=c:A$	$\mathbf{X}-!-B(a)=B(c):set$

Table 7.19 Substitution $b(x)$

Posit	Challenge	Defence
$\mathbf{X}-!-b(x):B(x)\ (x:A)$	$\mathbf{Y}-?_{b(x)-\text{subst}}-a=c:A$	$\mathbf{X}-!-b(a)=b(c):B(a)$

A.2.5.2   The Equality-Predicate

Table 7.20 Formation of the equality predicate

Posit	Challenge	Defence
$\mathbf{X}-!-\text{Id}(\dot{A},a,b):set$	$\mathbf{Y}\ ?^1_{\text{F-Id}}$	$\mathbf{X}\text{-}!\text{-}A:set$
	or	$a:A$
	$\mathbf{Y}\ ?^2_{\text{F-Id}}$	$b:A$
	or	
	$\mathbf{Y}\ ?^3_{\text{F-Id}}$	
	[the challenger has the choice]	

Table 7.21 From definitional equality to the equality-predicate

Posit	Challenge	Defence
$\mathbf{X}-!-\text{Id}(\dot{A},a,b):set$	$\mathbf{Y}\ ?^1_{\text{F-Id}}$	$\mathbf{X}\text{-}!\text{-}A:set$
	or	$a:A$
	$\mathbf{Y}\ ?^2_{\text{F-Id}}$	$b:A$
	or	
	$\mathbf{Y}\ ?^3_{\text{F-Id}}$	
	[the challenger has the choice]	

Table 7.22 Reflexivity for the equality predicate

Posit	Challenge	Defence
$\mathbf{X}{-}!-a:A$	$\mathbf{Y}$-? a-Id-$refl$	$\mathbf{X}$-!- r(a) : $Id(A, a, a)$
		Description: r(a) = a

Table 7.23 Substitution for the equality-predicate

Posit	Challenge	Defence
$\mathbf{X}$-!- p : $I(A, a, b)$	$\mathbf{Y}$-? $_{\text{pred.-susbst}}-b:A$	$\mathbf{X}$-!- q : $B(b)$
$\mathbf{X}$-! $a:A$		
$\mathbf{X}$-!- q : $B(a)$		

References

1. S. Magnier, *Approche dialogique de la dynamique épistémique et de la condition juridique* (College Publications, London, 2013)
2. M. Armgardt, *Das rechtlogische System der "Doctrina Conditionum" von Gottfried Wilhelm Leibniz*. Computer im Recht, vol. 12 (Elvert-Verlag, Marburg, 2001)
3. M. Armgardt, Zur Bedingungsdogmatik im klassischen römischen Recht und zu ihren Grundlagen in der stoischen Logik. Tijdschrift voor Rechtsgeschiedenis **76**, 219–235 (2008)
4. M. Armgardt, Zur Rückwirkung der Bedingung im klassischen römischen Recht und zum stoischen Determinismus. Tijdschrift voor Rechtsgeschiedenis **76**, 341–349 (2010)
5. A. Thiercelin, On two argumentative uses of the notion of uncertainty in Law in Leibniz's juridical dissertations about conditions, in [15, pp. 251–266]
6. A. Thiercelin, La Theorie Juridique Leibnizienne des Conditions. Ce que la logique fait au droit (ce que le droit fait à la logique). Ph.D.-Universite-Charles de Gaulle, Lille, 2009
7. A. Thiercelin, Epistemic and practical aspects of conditionals in Leibniz's legal theory of conditions, in *Approaches to Legal Rationality*, ed. by D. Gabbay, P. Canivez, S. Rahman, A. Thiercelin (LEUS-Springer, Dordrecht, 2010), pp. 203–216
8. W.G. Leibniz, *Sämtliche Schriften und Briefe*. Edited by the Deutsche Akademie der Wissenschaften zu Berlin. Darmstadt, 1923 sqq., Leipzig, 1938 sqq., Berlin, 1950 sqq. Cited by Series (Reihe) and Volume (Band)
9. J. Granström, *Treatise on Intuitionistic Type Theory* (Springer, Dordrecht, 2011)
10. H.-J. Koch, H. Rüßmann, *Juristische Begründungslehre* (Beck, München, 1982)
11. P. Boucher, *Leibniz: What Kind of Legal Rationalism?* in [15, pp. 231–49]
12. H. Schepers, Leibniz' Disputationen De Conditionibus: Ansätze zu einer juristischen Aussagenlogik. Studia Leibniziana, Supplementa **15**, 1–17 (1975)
13. E. Vargas, Contingent Propositions and Leibniz's Analysis of Juridical Dispositions, in [15, pp. 266–278]
14. M. Dascal, Leibniz's Two-pronged Dialectics, in [15, pp. 37–72]
15. M. Dascal (ed.), *Leibniz: What Kind of Rationalist?* (LEUS-Springer, Dordrecht, 2008)
16. S. Rahman, H. Rückert, New perspectives in dialogical logic. Synthese **1/2**, 127 (2001)
17. S. Rahman, H. Rückert, Dialogical connexive logic, in [16, pp. 105–139]
18. S. Rahman, J. Redmond, *Hugh MacColl et la Naissance du Pluralisme Logique* (College Publications, London, 2008)
19. C. Pizzi, T. Williamson, Strong Boethius' Thesis and Consequential Implication. J. Philos. Logic **26**, 569–588 (1997)
20. G. Priest, Negation as cancellation and connexive logic. Topoi **18**, 141–148 (1999)
21. H. Wansing, Connexive modal logic, in *Advances in Modal Logic*, vol. 5, ed. by R. Schmidt et al. (King's College Publications, London, 2005), pp. 367–383
22. H. Wansing. Connexive Logic, in *The Stanford Encyclopedia of Philosophy* (Spring 2014 Edition), ed. by Edward N. Zalta, http://plato.stanford.edu/archives/spr2014/entries/logic-connexive/

23. D. Steiner, Th. Studer, Total public announcements. Typoscript (2007)
24. M. Armgardt, Salvius Iulianus als Meister der stoischen Logik –zur Deutung von Iulian D. 34,5,13(14), 2–3, in *Liber amicorum Chris-toph Krampe zum 70. Geburtstag,* ed. by M. Armgardt, F. Klinck, I. Reichard, pp. 29–36. Freiburger Rechtsgeschichtliche Abhandlungen Bd. 68 (Berlin, 2013)
25. G. Sundholm, Constructions, proofs and the meaning of the logical constants. J. Philos. Logic **12**(2), 151–172 (1983)
26. G. Sundholm, Proof-theory and meaning, in *Handbook of Philosophical Logic,* ed. by D. Gabbay, F. Guenthner, vol. 3 (Reidel, Dordrecht, 1986), pp. 471–506
27. G. Sundholm, A plea for logical atavism, in *The Logica Yearbook 2000,* ed. by O. Majer (Filosofia, Prague, 2001), pp. 151–162
28. G. Sundholm, A century of judgment and inference: 1837–1936, in *The Development of Modern Logic,* ed. by L. Haaparanta (Oxford University Press, Oxford, 2009), pp. 263–317
29. G. Sundholm, Containment and variation; two strands in the development of analyticity from Aristotle to Martin-Löf, in *Judgement and the Epistemic Foundation of Logic,* ed. by M. van der Schaar (Springer, Dordrecht, 2013), pp. 23–35
30. J. Hintikka, *Knowledge and Belief: An Introduction to the Logic of the Two Notions* (Cornell University Press, Cornell, 1962)
31. A. Ranta, *Type-theoretical grammar* (Clarendon Press, Clarendon, 1994)
32. M. Armgardt, Presumptions and Conjectures in the Legal Theory of Leibniz. In this volume
33. D.M. Gabbay, J. Woods, Relevance in the law, in *Approaches to Legal Rationality,* ed. by D. Gabbay, P. Canivez, S. Rahman, A. Thiercelin (Springer, Dordrecht, 2012), pp. 239–264
34. G. Primiero, *Information and Knowledge. A Constructive Type-Theoretical Approach* (Springer, Dordrecht, 2008)
35. G. Primiero, Type-theoretical dynamics, in *The Realism-Antirealism Debate in the Age of Alternative Logics,* ed. by Rahman et al, (Springer, Dordrecht, 2012), pp. 191–212
36. J. Woods, *Paradox and Paraconsistency. Conflict Resolution in the Abstract Sciences* (Cambridge University Press, Cambridge, 2003)
37. D.M. Gabbay, J. Woods, Logic and the law. Crossing the lines of discipline, in *Approaches to Legal Rationality,* ed. by D. Gabbay, P. Canivez, S. Rahman, A. Thiercelin (Springer, Dordrecht, 2012), pp. 165–202
38. P. Martin-Löf, *Intuitionistic Type Theory – Notes by Giovanni Sambin of a Series of Lectures Given in Padua, June 1980* (Bibliopolis, Naples, 1984)
39. B. Nordström, K. Petersson, J.M. Smith, *Programming in Martin-Löf's Type Theory – An Introduction* (Oxford University Press, Oxford, 1990)
40. H. Rückert, *Dialogues as a dynamic framework for logic* (College Publications, London, 2011)
41. N. Clerbout, S. Rahman, *Linking Game-Theoretical Approaches with Constructive Type Theory: From Dialogical Strategies to CTT-Demonstrations* (Springer, Dordrecht, 2015, in print)
42. S. Rahman, N. Clerbout, L. Keiff, The constructive type theory and the dialogical turn: a new start for the Erlanger Konstruktivismus, in *Dialogische Logik,* ed. by J. Mittelstrass, (Mentis, Münster, 2015), pp. 127–184
43. P. Lorenzen, K. Lorenz, *Dialogische Logik* (Wissenschaftliche Buchgesellschaft, Darmstadt, 1978)
44. K. Lorenz, Basic objectives of dialogue logic in historical perspective, in [16: 255–263]
45. W. Felscher, Dialogues as a foundation for intuitionistic logic, in *Handbook of Philosophical Logic,* ed. by D. Gabbay, F. Guenthner, vol. 3 (Kluwer, Dordrecht, 1985), pp. 341–372
46. K. Lorenz, *Dialogischer Konstruktivismus* (De Gruyter, Berlin/New York, 2008)
47. K. Lorenz, *Philosophische Variationen: Gesammelte Aufsatze Unter Einschluss Gemeinsam Mit Jurgen Mittelstrass Geschriebener Arbeiten Zu Platon Und Leibniz* (De Gruyter, Berlin/New York, 2010)
48. K. Lorenz, *Logic, Language and Method – On Polarities in Human Experience* (De Gruyter, Berlin/New York, 2010)
49. S. Rahman, *Über Dialogue, Protologische Kategorien und andere Seltenheiten* (P. Lang, Frankfurt/Paris/New York, 1993)

50. S. Rahman, L. Keiff, On how to be a dialogician, in *Logic, Thought and Action*, ed. by D. Vanderveken (Kluwer, Dordrecht, 2005), pp. 359–408

51. L. Keiff, Dialogical Logic, in *The Stanford Encyclopedia of Philosophy* (Summer 2011 Edition), ed. by Edward N. Zalta. http://plato.stanford.edu/archives/sum2011/entries/logic-dialogical/

52. S. Rahman, Negation in the logic of first degree entailment and tonk: a dialogical study, in *The Realism-Antirealism Debate in the Age of Alternative Logics,* ed. by Rahman et al. (Springer, Dordrecht, 2012), pp. 213–250

53. N. Clerbout, Etude sur quelques sémantiques dialogiques. Concepts fondamentaux et éléments de metathéorie (College Publications, London, 2014)

54. N. Clerbout, First-order dialogical games and tableaux. J. Philos. Log. doi: 10.1007/s10992-013-9289-z. Online first publication

55. W. Kamlah, P. Lorenzen, *Logische Propädeutik*, 2nd edn. (Metzler, Stuttgart/Weimar, 1972)

56. W. Kamlah, P. Lorenzen, *Logical Propaedeutic*. English translation of [55] by H. Robinson, University Press of America, Lanham, 1984

57. P. Lorenzen, O. Schwemmer, *Konstruktive Logik Ethik und Wissenschaftstheorie*, 2nd edn. (Bibliographisches Institut, Mannheim, 1975)

58. J. Redmond, M. Fontaine, *How to play dialogues. an introduction to dialogical logic* (College Publications, London, 2011)

59. S. Rahman, L. Keiff, La Dialectique entre logique et rhétorique. Revue de Métaphysique et de Morale **66/2**, 149–178 (2010)

60. L. Keiff, Heuristique formelle et logiques modales non normales. Philosophia Scientiae **8/2**, 39–57 (2004)

61. L. Keiff, Introduction à la dialogique modale et hybride. Philosophia Scientiae **8/2**, 89–102 (2004)

62. L. Keiff, Le Pluralisme Dialogique. Approches dynamiques de l'argumentation formelle. Ph.D. thesis, Université de Lille, Lille, 2007

63. S. Rahman, A non normal logic for a wonderful world and more, in *The Age of Alternative Logics,* 2nd edn, ed. by van Benthem et al. (Kluwer-Springer, Dordrecht, Second edition, 2009), pp. 311–344

64. V. Fiutek, H. Rückert, S. Rahman, A dialogical semantics for Bonanno's system of belief revision, in *Constructions,* ed. by P. Bour et al. (College Publications, London, 2010), pp. 315–334

65. A. Ranta, Propositions as games as types. Synthese **76**, 377–395 (1988)

66. N. Clerbout, M.-H. Gorisse, S. Rahman, Context-sensitivity in Jain Philosophy. A dialogical study of Siddharsigani's commentary on the handbook of logic. J. Philos. Log. **40/5**, 633–662 (2011)

67. A. Popek, Logical dialogues from Middle Ages, in *Logic of Knowledge. Theory and Applications,* ed. by Barés et al. , (College Publications, London, 2011), pp. 223–244

68. S. Rahman, T. Tulenheimo, From games to dialogues and back: towards a general frame for validity, in *Games: Unifying Logic, Language and Philosophy*, ed. by O. Majer et al. , (Springer, Dordrecht, 2009), pp. 153–208

69. J. Redmond, *Logique dynamique de la fiction, Pour une approche dialogique* (College Publications, London, 2010)

70. M. Fontaine, Argumentation et engagement ontologique de l'acte intentionnel. Pour une réflexion critique sur l'identité dans les logiques intentionnelles explicites (College Publications, London, 2013)

71. H. van Ditmarsch, D. van der Hoek, B. Kooi, *Dynamic-Epistemic Logic* (Springer, Berlin, 2007)

72. J. Hintikka, *Logic, Language-Games and Information: Kantian Themes in the Philosophy of Logic* (Clarendon, Oxford, 1973)

73. S. Rahman, N. Clerbout, L. Keiff, On dialogues and natural deduction, in *Acts of Knowledge – History, Philosophy, Logic,* ed. by S. Rahman, G. Primiero (College Publications, London, 2009), pp. 301–336

Chapter 8
Legal Fictions, Assumptions and Comparisons

Giuliano Bacigalupo

Abstract Pierre Olivier distinguishes between two radically different conceptions of legal fictions: on the one hand, the conception of legal fiction developed by the commentators of the Middle Ages, which culminates in Bartolus's definition; on the other hand, the conception developed by the nineteenth century German scholar Gustav Demelius, who was followed, among others, by Joseph Esser. The main difference between the two approaches is individuated by Olivier in the fact that, while the former consider legal fictions as essentially implying an actual fictional element, the latter deny this. In other words, according to Demelius and those who follow him, the term "legal fiction" is a misnomer. In this article, I first provide an example of a legal fiction. In the second and third section, I rely on this example to analyze and assess the two competing accounts. Finally, in the fourth part, I advance a syncretistic account of legal fictions, which should thus point to a possible middle ground between the two competing positions. As it is often the case, there is probably some truth to both accounts; the problem – I will argue – is that both theories tell only a part and not the whole of the story. More precisely, it will be argued that legal fictions essentially involve the structure of "as if"-statements, and that the one-sidedness of the two competing accounts derives from the fact that one focuses too much on the "if"-component (the assumption), whereas the other focuses too much on the "as"-component (the comparison).

8.1 Introduction

Legal fictions are a powerful tool of law which may be traced back all the way to the Roman antiquity. One of the boldest examples may be found in the *lex corneliae*: if a Roman citizen was captured in war and died in slavery, he was deemed dead at the moment of enslavement so as to preserve the legal validity of his last will. Another very well known example is provided by the *fictio civitatis*: in order to enable a *peregrinus* (i.e. non-Roman citizen) to defend his interests in a civil court, he was

G. Bacigalupo (✉)
Fachbereich Rechtswissenschaft, Universität Konstanz, Konstanz, Germany
e-mail: giuliano.bacigalupo@uni-konstanz.de; giulianobacigalupo@googlemail.com

© Springer International Publishing Switzerland 2015
M. Armgardt et al. (eds.), *Past and Present Interactions in Legal Reasoning and Logic*, Logic, Argumentation & Reasoning 7, DOI 10.1007/978-3-319-16021-4_8

169

deemed a Roman citizen. Or more precisely, in such a trial, the judge had to pass his judgment as if the *peregrinus* were a Roman citizen ("ut si civis Romanus esset") (Gaius 4.37).

Even though very well acquainted with legal fictions, Roman law did not seem to feel the need to define this "curious artifice of legal reasoning" (see [7, p. 1]). This task will be left to the mediaeval glossators such as Cinus da Pistoia, Baldus de Ubaldis and Bartolus de Saxoferrato. All of them relied on the notion of a knowingly false assumption to capture the phenomenon of legal fictions. As it happens, it is their approach which hides behind most definitions of legal fictions that may be found in legal dictionaries up to these days.[1]

However, a renewed and polemic interest in the phenomenon of legal fictions in the nineteenth century led to a competing approach, which was developed by the German scholars Gustav Demelius [2], Edouard Hölder [6], August Sturm [10], Joseph Esser [4], and Joannes Eggens [3]. Demelius and those who followed him rejected the notion of a knowingly false assumption as a way to capture legal fictions and relied instead on the notion of comparison or, more precisely, equalization. To go back to the previous example of *fictio civitatis*, according to Demelius, we should not think that law is forcing us to assume what is not the case, namely that a *peregrinus* be a Roman citizen. Rather, what law is telling us is that the same rules which apply to Roman citizens in civil trials should be applied to the *peregrines*, too. In other words, the *peregrinus* and the Roman citizen are "equalized" with respect to civil trials.

In his monographic study on legal fictions, Pierre Olivier [8] brings the difference between the two competing approaches to the point: whereas the medieval glossators considered legal fictions to actually imply a fictional element, the German scholars in the tradition of Demelius deny this. To Demelius and those who follow him, the term "legal fiction" is a misnomer. It should be noticed, moreover, how Olivier argues for the standard medieval account of legal fictions: to him legal fictions really involve a fictional element, i.e. a knowingly false assumption.

My aim is to develop a middle ground solution that should enable us to vindicate some elements of both competing approaches: according to the approach I am going to defend, there is some truth in both the standard medieval approach and the revisionary modern one. The middle ground conception of legal fictions I will develop is indebted to an intuition that can be extracted from Hans Vaihinger's highly influential *Philosophie des Als Ob* [11]. From my perspective, Vaihinger's achievements lies not so much in his ambitious theory of fictions as omnipresent epistemological tools, but rather in his simple yet elegant logical interpretation of as-if statements as something which embeds both a conditional (an "if"-statement) and a comparison (an "as"-statement). Once "as if"-statements are broken down

---

[1]See for instance [5, p. 804]: "A legal fiction is a rule of law which assumes as true and will not allow to be disproved something which is false but not impossible".

into these two elements, it should become clear how the one sidedness of the two competing approaches to legal fictions lies in the fact that one focused too much on the "if"-part (the assumption), whereas the other focused too much on the "as"-part (the comparison).

8.2 An Example

Let us start by introducing an example that is not strictly speaking about laws, but pertains to the more general category of rules or norms. The advantage of this example is that it is as simple as one could possibly wish for and it does not require any background knowledge, neither legal nor historical, which may obfuscate the very general, logical point I am striving for. Consider a swimming pool with only two rules[2]:

(1) If and only if you are a woman, then you have access to the swimming pool.
(2) The lifeguard is deemed a woman.

The second rule is what legal scholars would term a legal fiction – of course in a legal context, but the difference between rules and laws is orthogonal to the present discussion. Evidently, the legal fiction (2) is introduced to allow the inference of (3) independently from the fact whether the lifeguard is a man or a woman[3]:

(3) The lifeguard has access to the swimming pool.

The problem, however, is that it is not entirely clear in which sense we can say that (3) follows from (1) and (2). Notice, moreover, that to term (2) a legal fiction does not provide any solution to the problem, but it is just a name for the problem – or perhaps even a misnomer. Indeed, if we substitute (2) with (2a), it is still not clear whether the inference should follow:

(2a) It is a legal fiction that the lifeguard is a woman.

In the next sections, I proceed in the following way. First, I substitute (2a) with Bartolus's definition of fiction and assess how effective such a definition is at yielding us the needed inference. Then, I proceed in the same way with Demelius's definition. Finally, I develop the above mentioned middle-ground solution and argue that it is the best suited one to make sense of the inference at stake.

---

[2]The example is due to Pfersmann [9].

[3]The debate whether inference-relations may hold between norms or rules – and not only between assertions – is a well-known subject of controversy. For a defence of the view that it does indeed make sense to speak of inference in the case of norms and rules, see [12] and [13].

8.3 Bartolus's Definition of Legal Fiction

According to Olivier [8, p. 17], Bartolus's definition represents the culmination of the mediaeval discussion on legal fictions. Moreover, since Olivier sides with the standard mediaeval account, it follows that to him Bartolus's definition is the one that should be retained by legal science, albeit a few minor modifications [8, p. 81]. The definition runs as follows:

(def. 1) A legal fiction is an assumption for legal purposes, about something which is certain, of something possible, not true but considered as true.[4]

The definition is a complex one and thus – following Olivier – it is helpful to break it down in its elements:

(a) Assumption (*assumptio*): As Olivier notices, this element is "the crux of the definition of legal fictions" since the notion of assumption is very close to the notion of fiction. Indeed, if one understands by fiction a conditional acceptance of the truth of a statement known to be false, and, on the other hand, one understands by assumption a conditional acceptance of the truth of a statement, then fictions turn out to be a kind of assumptions [8, p. 60].

(b) For legal purposes (*a iure facta*): According to Olivier, this element refers to the fact that the false assumption as well as its consequences are both permitted and prescribed by law [8, pp. 73–77]. This follows directly from the fact that legal fictions have to be laws, and so they lawfully permit and prescribe something. One way to look at this element could also be the following one: the *a iure facta*-element is what captures on the side of the *definiens* the meaning of the term "legal" on the side of the *definiendum*.

(c) Not true (*contra veritatem*): This element highlights the falsity of what is assumed. Once this element is added to the notion of assumption, it yields us the notion of fiction, at least as long as one considers a fiction to be a false assumption [8, pp. 62–69].

(d) Considered as true (*pro veritatem*): As Oliver notices, there seems to be something pleonastic in this characterization of the assumption as something considered to be true. Is it not the case that the notion of assumption analytically implies the notion of assuming something to be true? Olivier thus tries to make sense of this element as implying the non-rebuttable character of the assumption: the *pro veritatem*-element implies that the assumption cannot be challenged. A clear line may, thus, be drawn between legal fictions and legal rebuttable presumptions [8, p. 69].[5]

---

[4]The translation is mine. The original reads "Fictio est in re certa eius quod est possibile contra veritatem pro vertitate a iure facta assumptio" (cf. [8, p. 16]).

[5]A classic example of rebuttable presumption is the presumption of innocence: if this presumption could not be challenged, there would be of course no sense in having any (penal law) trial. The fact that legal fictions are not rebuttable is also stressed by the definition provided by *Black's Law Dictionary* (see above, footnote 1).

(e) About something which is certain (*in re certa*): Most plausibly, the *in re certa*-element means that the assumption at stake in legal fictions is about a fact which is certain. In the swimming pool scenario, for instance, we are indeed certain whether the lifeguard is really a man or a woman. But, as Olivier notices, if this is how we are supposed to understand this element, there seems to be again some redundancy between the *in re certa*-element and the *contra veritate* one: since the assumption is contrary to how things are, this seems to imply that we know – i.e. that we are certain – that things are otherwise [8, pp. 69–73]. However, one may notice that the *in re certa*-element has a clear epistemic dimension, so that one may indeed interpret it as making a positive contribution to the definition: not only the assumption is contrary to the truth, but we also have to know this. The legal fiction is a knowingly false assumption.

(f) Of something possible (*eius quod possibile*): Olivier notes how this element seems contradictory. Since we know that the assumption is – strictly speaking – false, to Olivier we are assuming something impossible [8, pp. 77–78]. But then how could the assumption be about something possible? However, it seems to me that Olivier is wrongly presupposing an epistemic notion of possibility: if I know that something is not the case, then it is epistemically impossible that it is the case. Instead, as Olivier himself acknowledges, Bartolus had a physical notion of possibility in mind: the legal fiction should imitate nature, and thus should not force on us a physically impossible assumption [8, p. 16]. In the case of the swimming pool scenario, this simply means that it should not be physically impossible for the lifeguard to be a woman.[6]

8.3.1 Bartolus's Definition at Work

Let us go back to the inference we are interested in. For starters, one may notice how not all the elements of Bartolus's definition of legal fiction need to be taken into consideration. In fact, some elements do not seem to be strictly relevant for the inference at stake but rather play the role of requirements that something has to fulfil in order for it to qualify as legal fiction.

First, it is a requirement that the legal fiction be a law (b). We may consider this requirement as fulfilled by (2), whereby of course we have to be careful to notice that (2) is not, strictly speaking, a law, but, more generally, a norm or a rule: It is a norm that the lifeguard is deemed a woman.

---

[6]One could argue that, in the case in which the lifeguard is a man, it would be physically impossible for him to be a woman. Thus, we would not be allowed to assume that the male lifeguard is a woman. To this line of reasoning I would object that it presupposes too strict a reading of the notion of possibility at stake. What Bartolus stresses is that it should not be physically impossible to assume that the lifeguard is a woman. The fact that the lifeguard is actually a man and, thus, that it may be physically necessary for it to be a man does not play any role here. In other words, the possibility should not be construed as a *de re* but rather as a *de dicto* possibility.

Second, it is a requirement that the assumption at stake in the legal fiction should not be true (c): Only in the case where the lifeguard is actually a man, we may consider (2) as a legal fiction. In fact, in all the cases were the lifeguard is a woman, we simply do not need (2).

Third, the epistemic element (e) also plays the function of a requirement: in order for (2) to be a legal fiction, not only it has to be false that the lifeguard is a woman, but we also have to know that it is false that the lifeguard is a woman.

Finally, the modal element, too, constitutes a requirement: for (2) to be a legal fiction, it has to be possible for the lifeguard to be a woman. And again, we may consider this requirement as fulfilled by (2).[7]

On the other hand, the elements of the definition which are crucial for the inference at stake seem to be (a) and (b): the *assumptio*-element and the *pro veritate*-element, of course interpreted as implying the non-rebuttable character of the assumption (otherwise, as addressed above, the *pro veritate*-element would be pleonastic). Thus, we may reformulate the inference we are interested in as follows:

(1) If and only if you are a woman, you have access to the swimming pool.
(2b) It is assumed and not-rebuttable that the lifeguard is a woman.
(3) The lifeguard has access to the swimming pool.

The question which interests us may now be formulated as follows: does (3) follow from (1) and (2b)?

A first, more general objection to the inference from (1) and (2b) to (3) is that law should not be allowed to alter reality: (2b) would literally force us to assume that the lifeguard is a woman even though we all know this not to be the case (of course, I am considering the situation in which the lifeguard is indeed a man). This gives rise to the classical objection levelled at the use of fictions in law, of which Jeremy Bentham has been the most forceful spokesman. In Bentham's vivid wording, "[legal fictions are] the most pernicious and basest form of lying" [1, p. 582].

However, it is not immediately evident why law should not be allowed to alter reality or to lie. This is especially the case since legal fictions do not mean to deceive anyone, but rather are means for achieving a certain legal result, e.g. that the male lifeguard has access to the swimming pool.

A second, less ethical and more logical objection would be to point to the fact that (3) does not strictly follow from (1) and (2b): Since one of the premises is in the scope of a non-rebuttable assumption, so the conclusion does have to be in the scope of a non-rebuttable assumption, too.[8] Strictly speaking, form (1) and (2b) we cannot infer (3) but only (3a):

(3a) It is assumed and not-rebuttable that the lifeguard has access to the swimming pool.

---

[7]See previous footnote.

[8]Of course, this is only true under the rather uncontroversial premise that assumptions are closed under entailment.

But one may argue that the difference between (3) and (3a) may be played down in the case of laws, or, more generally, of norms or rules. If we may infer the rule that the lifeguard has access to the swimming pool, or if we can only infer the rule that it is assumed that the lifeguard has access to the swimming pool, there does not seem to be too big a difference. The normative effect seems to be the same, namely that the lifeguard has access to the swimming pool even though it is a man. This is even more the case since the assumption cannot be challenged: it is a non-rebuttable assumption.

A third objection may point to another logical inconvenient of Bartolus's definition: if (2b) is an assumption that cannot be questioned, then we have to accept all the consequences that follow from it. Yet we may want to avoid this in a legal or rule-based context. For instance, I assume that the swimming pool also has the following non-bikini friendly rule:

(4) If you are a woman, you must wear a one-piece swimsuit.

Do we want to force a male lifeguard to wear a one-piece swimsuit? Under (def. 1) someone may argue for it and a judge or, perhaps more aptly, the responsible for the swimming pool, would have to concede the argument.[9]

This problem, however, may be addressed by requiring legal fictions to specify their field of application. For instance, in cases where we want the legal fictions to be applied only with respect to certain norms, this has to be explicitly mentioned by the legal fiction. For instance, one may have to reformulate (2b) as follows:

(2b′) With respect to rule (1), it is assumed and non-rebuttable that the lifeguard is a woman.

Like this, we may rely on rule (2b′) to infer (3) or more precisely (3a), but we would not be able to rely on (2b′) and (4) to infer that the lifeguard has to wear a one-piece swimsuit even though he is a man.

To sum up, there seem to be at least three objections that may be raised against (def. 1). However, none of them may be deemed as inflicting a knock-down blow to Bartolus's definition: the first ethical objection and the second and third logical ones may all very well be addressed without having to alter in any essential way (def. 1). Still, a definition of legal fictions that would avoid such problems altogether would be preferable. This is especially true of the second objection, which leads to a gerrymandering of the conclusion from (3) to (3a).

---

[9]Less facetious examples show how the problem is indeed a serious one. One just need to think of the fiction of corporations as persons: if corporations really were persons, we would have to grant them a right to vote, a right to freedom of speech, and even a right to freedom of religion. The second and the third really happened in the Supreme Court's law cases "Citizen United v. Federal Election Commission" (2010) and "Sebelius v. Hobby Lobby Stores, Inc." (2014).

8.4 Demelius's Definition of Legal Fiction

Let us turn to the alternative approach, which denies that legal fictions really involve a fictional element. Demelius's definition of legal fictions, as quoted by Olivier [8, p. 50], is not a concise one:

(def. 2) [Legal fictions are] legal norms, by which a factual relationship, by equating it to another factual relationship which has already been juristically ordered, is ordered according to law, and by which it is equated as regards its juristic nature and consequences to the example and treated in the same manner.

It is important to underline how this definition does not include any notion that can be, in one way or the other, related to the notion of fiction. This does not happen by chance, since according to Demelius legal fictions are not, in any sense of the word, fictions. Or, as Olivier puts it, "Demelius regarded the fiction not as a method of thought or reasoning, but as a form of expression" [8, p. 41].

Besides Demelius, other proponents of the view that legal fictions are not really fictions but just a "form of expression" are Hölder [6], Sturm [10], Esser [4] and Eggens [3] (cf. [8, pp. 50–5]). Even though one cannot say that all these authors subscribe to Demelius's definition, there is a notion which runs through all their approaches, namely that of equalization (in German, *Gleichstellung*). According to them, the notion of equalization should take the place of the notion of assumption, which was the *crux* of Bartolus's definition. In short, what these authors all agree upon is that a rule such as (2) should not be interpreted as implying an assumption contrary to how things are. To them, the real meaning of the rule that the lifeguard is deemed a woman consists in the fact that the rules that apply to women should also be applied to the lifeguard, no matter whether he is a man or a woman.

Olivier also provides the reason why Demelius and the authors who followed in his footsteps have an aversion to the alleged presence of actual fictions in law. Laws – according to them – do not state anything about reality, since the only thing they do is to prescribe courses for human conduct. Thus, since fictions are allegedly assumptions contrary to reality, it cannot be the case that law contains fictions. However, according to Olivier, who – it will be remembered – sides with Bartolus's definition, the just sketched line of reasoning is fallacious. Laws – Olivier notices – do not exist *in vacuo*, since they are brought into operation by the occurrence of certain facts. As he writes [8, p. 55]:

In its simplest form the law can always be expressed in conditional form: if the facts are so and so, then the legal result is such and such.

Since the antecedent of the conditional to which every law may be reduced is a statement about how things are, there clearly is a place for fictions in law. More precisely: "the fiction changes the reality in front of the court" [8, p. 56]. I will come back to these important remarks in the fifth section.

8.4.1   Demelius's Definition of Legal Fictions at Work

As we have seen, to Demelius and his followers a legal fiction is just an equalization for the purposes of law. From this perspective, the inference that interests us would take the following form:

(1) If and only if you are a woman, you have access to the swimming pool.
(2c) The lifeguard is like a woman.
(3) The lifeguard has access to the swimming pool.

Again, the question we should address is whether (3) follows from (1) and (2c), and, thus, whether the interpretation of legal fictions as equalization is convincing.

The first problem we are confronted with is that it is not at all clear how we should treat inferences that rely on likeness. More precisely, from a strictly logical point of view, similes do not prove anything. Take for instance the famous Homeric simile that Achilles is like a lion. What can we infer from it? It is true that lions are strong; thus, one is tempted to infer that Achilles, too, is strong. Or, according a certain anthropomorphic view, lions are brave, so that one may think that we are entitled to infer that Achilles, too, is brave. The problem, however, is that logically, a statements such as "Achille is like a lion" simply says that for *some* properties, if a lion instantiate this property, so does Achilles. But which are these properties, we cannot really tell. This is the vagueness of similes.

However, the situation is perhaps not so desperate in the case at hand. Demelius's definition of legal fictions does not reduce them to similes but to something more robust: legal fictions do not tell us that *some* legal consequences which apply to one case should be applied to another. Instead, the definition says that all legal consequences that apply to one case should be applied to another case. Equalization is more than just likeness.

From this perspective, we have to substitute (2c) with (2d):

(2d) For all the legal consequences, if these legal consequences apply to women, then they apply to the lifeguard, too.

Now the question should be whether (1) together with (2d) yield us the needed inference to (3). However, as soon as this question is raised, it becomes clear how we have solved one problem only to end up with a much more dramatic one. If we take the crucial case into consideration in which the lifeguard is a man, then (1) would force us to infer that the lifeguard does not have access to the swimming pool, whereas (2d) would force us to infer that the lifeguard does have access to the swimming pool. But legal fictions clearly cannot be deemed to make a system of norms inconsistent. Thus, we have to conclude that Demelius's definition cannot capture the notion of legal fiction.

8.5   Legal Fictions as Second-Order Rules

The upshot of the two proceeding sections is that the mediaeval approach to legal fictions is much more effective at yielding us the needed inference then the modern one. From this perspective, we have thus to side with Olivier. The only caveat was that some objections may be raised against the mediaeval account. None of these objections, however, qualified as full blown refutation. Nevertheless, as already noted, a view of legal fictions that would avoid (at least some of) these objections and the adjustments required by them would be preferable. This is the aim of the present section: to develop an account of legal fictions that enables us a straightforward and unproblematic inference to (3). In order to do this, I will both build upon the key notion of assumption at stake in the mediaeval account and vindicate the key notion of equalization at play in the modern account. The two key-notions of both approaches will thus join forces to cash out the phenomenon of legal fictions. Yet both the assumption and the equalization at stake in this new definition will not exactly match the kind of assumption and comparison implied by Bartolus and Demelius.

Let me start with some preliminary remarks. I fully subscribe to Olivier view that laws or for that matter norms or rules do not exist in a vacuum but have to be anchored to reality. Moreover, I also agree that this insight can be brought to light by the fact that laws may all be formulated as if-statements of the following form: if things are so and so, then the legal result is such and such. Or, in other words, we should apply law in relation to how things actually are, so that how things are determines the legal result. Yet – and this is the crucial part – the problem with legal fictions is that they clearly infringe this rule: a legal fiction says that, in certain circumstances, laws should not yield us a legal result in relation to how things actually are, but should yield us the same legal result *as if* things were different from what they are.

Let me explain the insight I have just sketched. Consider for a moment the situation in which the swimming pool has only one rule, namely that if and only if you are a woman, then you have access to the swimming pool (1). Evidently, this rule is intended to be applied in relation to how things actually are. To wit, if someone is a man, he does not have access to the swimming pool, and if someone is a woman, she has access to the swimming pool. But the introduction of the second rule, namely that the lifeguard is deemed a women (2), clearly introduces an exception to the unspoken rule: rule (2) says that if someone is the lifeguard, the rules should not be applied with respect to how things actually are, but rather *as if* the lifeguard were a woman. From this perspective, (2) is a meta-rule or a rule of second-order, which tells us how the other rules should be applied to a specific case.

Schematically, the situation may be spelled out as follows. In every rule-system we have an unspoken meta-rule, or a rule of second-order:

(0) Rules yield us normative consequences in relation to how things are.

However, as soon as a fiction makes its way in a system of rules, this implies a modification of the implicit rule (0):

(0a) If not stated otherwise, rules yield us normative consequences in relation to
how things are.

According to this perspective, also legal, or more generally, normative fictions
have to be understood as second-order rules:

(def. 3)   A normative fiction is a rule that says that if things are such and such, then
all first order rules have the same normative consequence as if things were
so and so.

In what follows, I will explore whether this third definition is more effective than
the previous ones at validating the inference at stake in the example of the swimming
pool. In order to achieve this, it will be crucial to provide an interpretation of what
has become the new crux of the definition, i.e. the "as if"-particle.

8.5.1   The Definition of Legal Fiction as a Meta-Law at Work

According to the interpretation of legal fictions as rules of second-order, the
inference that interests us takes the following shape (notice that we also have to
introduce the unspoken rule (0a)):

(0a) If not stated otherwise, rules yield us normative consequences in relation to
how things are.
 (1) If and only if you are a woman, you have access to the swimming pool.
(2d) If someone is the lifeguard, then all first order rules have the same normative
consequences as if the lifeguard were a woman.
 (3) The lifeguard has access to the swimming pool.

Intuitively, this inference follows – or at least I would argue so. The problem,
however, is that even though we all very often rely on "as if"-statements, both in
scientific and not scientific contexts, we rarely spend time considering what the
actual meaning of this particle is and, even more crucially, which are the inferences
allowed by it. Thus, before being in the position to assess the inference to (3), a brief
excursus on "as if"-statements will be necessary.

8.5.1.1   Vaihinger's Analysis of "as if"-Statements

As announced in the introduction, in order to analyze "as if"-statements I rely
on a very simple but compelling intuition that can be found in Hans Vaihinger's
Philosophie des Als Ob. According to Vaihinger, "as if"-statements are statements
which embed both a conditional (what is usually introduced by the particle "if")
and a comparison (what is usually introduced by the particle "as"). More precisely,
"as if"-statements are statements which compare what is actually the case with the
consequence of something that is not actually the case (cf. [11, p. 584–91]). To fully
understand Vaihinger's intuition, it is helpful to consider his discussion of another
Homeric example:

(5) Hector fell as if he was an oak tree

To Vaihinger, the full thought behind Homer's verse is the following:

(5a) Hector fell as he would fall if he was an oak tree.

This reformulation brings to the fore how the particle "as" introduces a comparison or an analogy, whereas the particle "if" introduces a conditional statement. This conditional statement, moreover, is characterized by a specific trait: the antecedent of the conditional does not have to be satisfied in reality (it is in the subjunctive form). As contemporary philosophers are keen to say, the conditional is a counterfactual one. If we put the two things together, we indeed see how an "as if"-statement is a comparison between what is actually the case and a consequence of what may not actually be the case. We all know that Hector is not an oak tree, but if he happened to be an oak tree, he would fall in a similar way as he actually fell.

However providing us with a crucial insight, there is a problem to which Vaihinger did not pay enough attention: it is not all clear which inferences are validated by his interpretation of "as if"-statements. In addition, this problem becomes particularly acute if we want to put the same construction to work in a scientific context – which, as it happens, was Vaihinger's actual aim. For instance, let us say that we all agree that if Hector were an oak tree, he would fall very heavily. May we infer from this that Hector actually fell very heavily? Clearly not, since, as long as we interpret the particle "as" as simply implying a comparison or similarity, this would mean that only for some properties that hold in the counterfactual scenario, this property also has to hold in the actual scenario. And clearly we cannot tell whether the property of falling heavily is within this set of properties. Once more, we are confronted with the vagueness of similes.

My suggestion would thus be that, once the "as if"-construction is employed in a scientific context, we should not consider the particle "as" as a simple comparison, but rather – relying on Demelius's intuition – as an equalization: all the properties that hold in the counterfactual scenario also hold in actuality. This trait was *de facto* already introduced in (def. 3), which states that *all* rules have the same consequences, as if things were such and such.

Once this regimentation of the "as if"-construction is put into place, it is clear how (1) and (2d) allows us to infer (3) without further ado: since we all agree that if the lifeguard were a woman, the lifeguard would have access to the swimming pool (by 1), and since the interpretation of normative fictions provided by (3) says that every legal consequence that applies to the counterfactual scenario also applies in actuality, we may conclude that the lifeguard actually has access to the swimming pool.[10]

---

[10]Note how the same regimentation would be highly questionable in the literary context: we clearly do not want that Hector actually has all the properties that he would have if he would be a falling tree. This would lead to paradoxical inferences, as for instance that Hector is made of wood or has branches (I am indebted to Shahid Rahman for this remark).

8.5.2 The Meta-Law Definition of Legal Fiction as a Middle Ground Solution

As just spelled out, the definition of normative fictions as meta-rules presents the advantage that it validates the crucial inference without any need to gerrymander the conclusion. More precisely, there is no need to put the conclusion within the scope of an assumption, as was the case with Bartolus's definition. But the point I would also like to stress is that this approach to normative fictions vindicates to a certain extent intuitions that can be found both in Bartolus's fictionalist view of legal fictions as assumptions, and in Demelius's non-fictionalist view of legal fictions as comparisons, or, more precisely, as equalizations.

The view that legal fictions are essentially linked to assumptions, i.e. (def. 1), is vindicated to the extent that legal fictions do involve assumptions, even though at the meta-rule level. Legal fictions do not force us to assume what is not actually the case in a non-rebuttable way. Instead, legal fictions tell us that, in order to know how laws should be brought to bear in a specific case, we should consider a counterfactual scenario and not how things are. Notice, moreover, that in this sense we are dealing with a perfectly ordinary case of assumption: We are dealing with a theoretical assumption, i.e. an assumption that enables us to explore the legal consequences of something. To be more precise, one may want to avoid the talk of "theoretical assumption", since after all we are dealing with norms. So, it may be more accurate if we refer to this kind of assumptions as normative assumptions: in order to know how we should apply rules in a given scenario, we should not look at how things actually are but rather assume a counterfactual scenario.

Meanwhile, also the view that legal fictions are essentially linked to a comparison or, more precisely, an equalization, i.e. (def. 2), is vindicated, even though one more time this happens only at the meta-rule level. Normative fictions tell us that in a given circumstance rules should be applied in exactly the same way as they would be applied if things were so-and-so. Thus, what we are confronted with is an equalization of the normative consequences of a rule according to the assumption that things are in a certain way with the actual normative consequences of the rule in a given circumstance.

Finally, we may raise the question as to what extent the label of normative fiction is appropriate. According to the approach just defended, there is indeed a fictional element in normative fictions, namely what is expressed by the conditional implied in the "as if"-construction: since it is a subjunctive conditional, it is a conditional that considers things which may be different from how they actually are (note however that they do not have to be different). To go back to the swimming pool example, the subjunctive conditional tells us to assume a situation in which it is the case that the lifeguard is a woman, whereby it may actually be the case that the lifeguard is not actually a woman, and see what consequences follow from this assumption. And since the notion of a conditional assumption that may be contrary to how things actually are captures the notion of fiction, we may see how it is not a misnomer to label normative fictions as fictions.

At the same time, it should be stressed how normative fictions are not just a fiction. Fictionality is – so to speak – only one element of normative fictions, which – by far – does not cover the whole story that there is to tell about them. While reading *War and Peace*, for instance, we conditionally assume something that may be different from the actual course of events (note, however, that it does not have to be so, as every reader of *War and Peace* will know). Moreover, one may also want to say that, without entering into problems of theory of literature or more generally philosophy of art, there is something theoretical to this assumption: to entertain the thought that the events in *War and Peace* really happened enables us to understand something – to put it crudely and somewhat naively – about human nature. But this is already the end of the story there is to tell about the more general notion of fiction.

With normative fictions things are essentially different: we do have a fictional element, but this fictional element has to be complemented by the equalization element in order for us to acquire the full picture of normative fictions. Normative fictions are in this sense much more real than aesthetic fictions.

8.5.3   Objections

Despite its effectiveness at validating the needed inference, the meta-rule view of normative fictions may also be targeted by objections. First, the moral objection that law should not lie may be adjusted so as to address the approach to normative fictions as second-order rules. In this case, one may simply have to argue that there should be no exception to the general rule that normative consequences should only be attached to how things actually are.

However, as soon as one considers that normative fictions are not entirely disconnected from how things actually are, this objection looses most of its force. Normative fictions simply say that, given a certain condition, which of course has to apply in the actual world – as for instance with the condition that someone is the lifeguard –, some normative consequences have to be attached to this condition. True, in order to find out which are these consequences, one has to consider a counterfactual situation. But this does not mean that – to use Olivier's expression – normative fictions exist in a vacuum. Instead, normative fictions are strongly rooted in reality.

A second and more problematic objection is that normative fictions may validate too many inferences. In fact, the same problem that confronted Bartolus's definition of fiction may also challenge the conception of fictions as meta-rules: If we add the rule (4) to our example of the swimming pool, the normative fiction about the lifeguard (2d) would validate the inference that the lifeguard, too, should wear a one-piece swimsuit.

This second objection may be easily addressed following the same strategy as before: if necessary, legal fictions should specify with respect to which other rule or rules the legal fiction should be applied. This means that we should move one more step from (2d) to (2d′):

(2d′) If someone is the lifeguard, then the first order rule (1) has the same normative consequences as if this someone were a woman.

This of course would also imply a minor change in the definition of normative fiction given above. We would have to move from (def. 3) to (def. 3a):

(def. 3a) A normative fiction is a rule that says that if things are such and such, then all (or some given) first order rules have the same normative consequence as if things were so and so.

Sometimes legal fictions may *de facto* not require any kind of restriction to the first-order rules for which they are relevant. But in most cases the legislator would be well advised to introduce such a restriction. The reasons are mainly two: First, it is only with respect to some specific first-order rule or set thereof that the normative fictions are introduced; Second, since legal systems undergo changes, even though a restriction may not be needed at time t_0, it may always be needed at time t_1.

A last objection may be that the given interpretation of legal fictions flies off in the face of (certain kinds of) essentialism: an essentialist about the property of being a woman or being a man, for instance, would deny that we may consider the counterfactual scenario in which the actually male lifeguard is a woman.[11] For according to him there simply is no such counterfactual scenario: a man will always be a man. Thus, no inference may be drawn from a non-existing counterfactual scenario.

However, this objection falls short of offering a refutation of the attempted definition of legal fiction. A first reply may simply point to the fact that this objection builds upon the assumption that the counterfactual scenario should be construed as a *de re* possibility: It is of the individual who actually is a man that we are saying that he could be a woman. But nothing compels us to give such a reading of the possibility at stake, since a *de dicto* possibility would do just as well. What we need is just that the statement "the lifeguard is a woman" is possibly true.[12]

A second reply may point to the fact, that even if (at least some) legal fictions had to be construed so as to imply a *de re* possibility, this would simply lead to some restrictions on the kind of fictions that may be introduced in a legal system. This, however, would be very different from a refutation of the definition of legal fictions at stake. These restrictions would then be more or less severe in relation to the kind of essentialism at stake. Be that as it may, it would be difficult for any essentialist to argue that we cannot consider a counterfactual scenario in which a *peregrinus* becomes a Roman citizen (*fictio civitatis*) or someone died earlier than when he actually died (*lex corneliae*). True, I have to concede that the example of

---

[11]Of course, being a man and being a woman should be defined as having the relevant kind of DNA. Otherwise, the essentialist about sex would be refuted by the simple fact that we have sex reassignment surgery.

[12]Cf. above, footnote 6.

fiction upon which I relied throughout this study is a rather bold one, perhaps even bolder than the *lex corneliae*. But then again, I have to admit that I do not know if there ever was a swimming pool with such a curious norm.

8.6 Conclusion

In this article, I have advanced a definition of legal fictions as second-order laws which essentially involve an "as if"-formulation. More precisely, since the discussion was more general in character and did not consider any specific trait that distinguishes laws from the more general category of rules or norms, what has been defended is a more general definition of normative fictions as second-order "as if"-rules. Crucial to the analysis of normative fictions was the interpretation of "as if"-statements as implying both a conditional, and more precisely a subjunctive conditional, as well as an analogy, or, more precisely, equalization. From this perspective, the two alternative approaches to legal fictions that have been analyzed by Olivier are to a certain extent correct: both the assumption crucial to Bartolus's definition and the equalization crucial to Demelius's definition do really play a role in legal fictions and, more generally, in normative fictions. What the two alternative approaches failed to notice, however, was that they were only telling one part of the story, while at the same time missing a crucial element, namely that legal fictions are rules of second order, i.e. rules that say how (some given) first-order rules should be applied to specific circumstances.

Acknowledgments This work has been supported by the DFG ANR Project JuriLog (ANR11 FRAL 003 01). I would like to thank the following members of the project for their helpful comments and feed-back: Matthias Armgardt, Reinhart Bengez, Karlheinz Hülser, Bettine Jankowski, Sébastien Magnier, Shahid Rahman, Juliette Sénéchal, and Juliele Sievers.

References

1. J. Bentham, *The Works of Jeremy Bentham, Published under the Superintendence of his Executor, John Bowring,* 11 vols. (William Tait, Edinburgh), pp. 1838–1843.
2. G. Demelius, *Die Rechtsfiktionen in ihrer geschichtlichen und dogmatischen Bedeutung* (Böhlau, Weimer, 1858)
3. J. Eggens, *Over het fingeren van Rechtsficties* (Erven F. Bohn, Amsterdam, 1958)
4. J. Esser, *Wertund Bedeutung der Rechtsfiktionen* (Klostermann, Frankfurt am Main, 1940)
5. B.A. Garner (ed.), *Black's Law Dictionary,* 5th edn. (West Publishing Company, St. Paul, 1979)
6. E. Hölder, Die Einheit der Correalobligation und die Bedeutung juristischer Fiktionen. Archiv für die Civilistische Praxis **69**, 203–240 (1886)
7. N.J. Knauer, Legal fictions and juristic truth. St. Thomas Law Rev. **23**, 1–48 (2010)
8. P. Olivier, *Legal Fictions in Practice and Legal Science* (Rotterdam University Press, Rotterdam, 1975)

9. O. Pfersmann, Les modes de la fiction en droit et en littérature, in *Usages et théories de la fiction*, ed. by F. Lavocat (Presses Universitaires de Rennes, Rennes, 2004), pp. 39–61
10. A. Sturm, *Fiktion und Vergleich in der Rechtswissenschasft* (Helwing, Hannover, 1915)
11. H. Vaihinger, *Philosophie des Als Ob. System der theoretischen, praktischen und religiösen Fiktionen der Menschheit auf Grund eines idealistischen Positivismus. Mit einem Anhang über Kant und Nietzsche* (Felix Meiner, Leipzig, 1911)
12. P. Vranas, New foundations for imperative logic I: Logical connectives, consistency, and quantifiers. Noûs **42**, 529–572 (2008)
13. P. Vranas, In defense of imperative inference. J. Philos. Log. **39**, 59–71 (2010)

Chapter 9
Reasoning with Form and Content

Juliele Maria Sievers and Sébastien Magnier

Abstract In this paper, we propose a logical frame allowing us to display the argumentative features behind legal decisions. This undertaking is motivated by Hans Kelsen's solution to a well-known puzzle in legal philosophy, called Jørgensen's dilemma. In the first part of the text, we deliver a detailed presentation of the problem, as well as two of the many attempts to a solution. Then we present what we consider to be the final solution, given by the legal philosopher Hans Kelsen. Based on this approach, the second part presents our attempts to provide, by means of a dialogical frame, an original application of the kelsenian solution in the field of legal justification. This logical frame not only perfectly displays Kelsen's approach but it also allows to express, debate and justify the legal reasoning without transgressing the limits between the legal field of normative creation and the scientific field of normative justification.

9.1 Introduction

The closeness between the domains of law and philosophy is corroborated by the prolific discussions concerning the foundations of the theory of law, namely the questions involving the definition of a legal norm. In a wide sense, norms are orders, commands deriving from human will; but legal norms specifically are obligatory imperatives, binding prescriptions, they are "ought" (*Sollen*) statements. According to Kelsen, there is a "methodological abyss", a gap between the "Ought" (*Sollen*) domain, where we find the norms, and the "Is" (*Sein*) domain, where we find

J.M. Sievers (✉) • S. Magnier
University of Lille 3, UMR 8163 Savoirs, Textes, Langage, 3 Rue du Barreau, 59650 Villeneuve-d'Ascq, France
e-mail: juliele.sievers@univ-lille3.fr; sebastien.magnier@bbox.fr

© Springer International Publishing Switzerland 2015   187
M. Armgardt et al. (eds.), *Past and Present Interactions in Legal Reasoning and Logic*, Logic, Argumentation & Reasoning 7, DOI 10.1007/978-3-319-16021-4_9

the sentences. Jørgensen's dilemma puts in question which kinds of interaction seem to prevail in the relation between prescriptions or imperatives (*Sollsätze*) and descriptions or sentences (*Sätze*).[1]

This dilemma, as pointed out by Giovanni Sartor, "consists in the supposed necessity of making the following choice: Either one rejects using logic in the law (and more generally, in practical reasoning) or one has to admit that legal contents (practical noemata) can be true or false".[2] In other words, the dilemma deals with the question whether if imperatives can be a part of a logical inference, as one of the premises or as the conclusion. If a first reaction points to a negative answer (since norms cannot be true or false[3]), most of us will have to agree that inferences involving norms make part of our daily lives and that the evidence of their legal validity[4] seems to be out of question. This difficulty in understanding the role of imperatives in our reasoning about norms intrigued also Hans Kelsen, who severely criticized the answers previously given.

First of all, our aim is to explain how Kelsen deals with the dilemma, by arguing that there cannot be any logical derivation between imperatives and sentences. Then, we will show how Kelsen's approach offers a theoretical frame to a brand new treatment of problems like Jørgensen's dilemma and consequently helps us to provide an original answer to the first problem depicted in [5] (i.e.: "How can deontic logic be reconstructed in accord with the philosophical position that norms are neither true nor false?").

In order to do so, we reconstruct the relations between law and legal science using a particular approach of logic, namely the dialogical approach. Not only this framework allows us to respect the methodological abyss pointed out by Kelsen, but it also displays the interactions which are inherent to the legal reasoning. Our attempts are particularly directed to aspects concerning the justification of legal decisions, which will be considered in a logical frame, putting in evidence the argumentative process involved in the legal reasoning.

9.2   Understanding Jørgensen's Dilemma

Jørgensen's dilemma is inserted in the discussions concerning normative reasoning and deontic logic. It states the problematic fact that, even if it is commonly accepted that norms cannot receive logical treatment since they are not true or false, it seems

---

[1]This distinction will also be important in the differentiation between legal norms (*Rechts-Norm*) and normative propositions (*Rechts-Satz*).

[2]See [17, chap. 15, p. 420].

[3]Norms are not facts, they are the "product" or the sense of an act of will, they are wanted by someone, and directed towards someone else's behaviour.

[4]Here legal *validity* means the specific existence of a norm in a legal order. For a norm to be valid or existent in the legal order means that all the legal procedures to its creation were respected.

to be legitimate to derive a specific norm from a general one, through a rational operation such as a syllogism. This syllogism should then be called "practical", because it contains Norms as the major premise and as the conclusion.

The existence of practical syllogisms would affect many aspects of our understanding, dealing with questions going from the rationality of the normative discourse to the justification of moral acts and decisions, crossing controversial aspects such as the conflicts between norms, moral dilemmas and the possibility of a logic of norms.

9.2.1   First Attempts: Jørgensen and Ross

The easiest way to understand this problem, first announced by Jørgen Jørgensen in 1937, and named after this author by Alf Ross some years later (1944), is to immediately invoke some examples.

In our daily lives, we are frequently confronted with legal norms and imperatives in general, and we are demanded to ratiocinate and theorize about them. Most frequently, we do so without paying attention (see Jørgensen's classical example in Table 9.1).

The question imposed considers whether imperatives can be a part of a logical inference, as a premise or as the conclusion. Since norms cannot be true or false, it would seem reasonable to answer negatively. However, most of us will agree that practical syllogisms of the form of the example above seem to be legitimate, even if we have no criteria to distinguish valid practical syllogisms from invalid or arbitrary ones. This difficulty in understanding the role of imperatives in our reasoning about norms, commands, orders and in the legal argumentation was called by Alf Ross the "Jørgensen's Dilemma".

Let us analyse some examples given by Jørgensen and Alf Ross, indicating the fact that the notion of practical syllogism is misleading and erroneous. Following a kelsenian approach, the primal obstacle to the logical treatment of imperatives in a practical syllogism lies on the fact that they are not the objective sense of an act of will.

9.2.1.1   Jørgensen's Answer to the Dilemma

The main problem concerning Jørgensen's dilemma does not concern the fact of deriving norms from facts (or indicative sentences), but consist of logically derive a

Table 9.1 Jørgensen's classical example

| Keep your promises |
| This is a promise of yours |
| Keep this promise |

norm from another norm.[5] His first reaction to this problem is to say that imperatives (norms) cannot be a part of a syllogism, because they cannot be true or false. But then he proposes another approach to the logical treatment of those imperatives: Jørgensen says that the answer to the conundrum lays in the difference between the imperative factor and the indicative factor present in the norm.

Since there is no imperative without "something" which is demanded to be performed, there is always a possibility to detach the object of demand from the imperative. In this way, the *imperative* factor concerns the expression of the subject's state of mind: the act of commanding, the act of giving an order. The *indicative* factor concerns the specific content of the norm and has a propositional feature, thus it may be treated with some logical tools.[6]

In order to be able to logically manipulate the imperative we must, first of all, subtract the indicative factor from the imperative. So, from "Tax evasion must be punished by fine and/or imprisonment" we subtract the indicative sentence "The norm demands that 'Tax evasion is punished by fine and/or imprisonment'" and, in a simpler way, we can directly convert it to the indicative sentence "Tax evasion is punished by fine and/or imprisonment". Now, since evidence shows that Jérôme C. practiced tax evasion, we can derive another indicative sentence, namely, "Jérôme C. is punished by fine and/or imprisonment". From that, we go the same way back by adding the imperative factor and reformulating a norm such as "Jérôme C. must be punished by fine and/or imprisonment".

To Jørgensen, the translation procedure explains the efficacy of the practical syllogism, in the same way as it justifies the inference process. The syllogism is legitimate since, in fact, the inference is made concerning the indicative factor, and the norm is obtained by the reconstruction of the imperative factor.

9.2.1.2 Ross' Answer to the Dilemma

Ross severely criticizes Jørgensen's answer. The problem, says Ross, is that we are not able to reconstruct the norm "Jérôme C. must be punished by fine and/or imprisonment" (I_2 in the Fig. 9.1 below) from the first norm "Tax evasion must be punished by fine and/or imprisonment" (I_1 in the Fig. 9.1). Jørgensen's logic of satisfaction is useless to explain how we derive new norms as conclusions in the practical syllogisms.[7] In the Fig. 9.1, Ross illustrates the gap through this illustration, where "I" stands for "Imperative" and "S" stands for "Sentence".

So, concerning the dilemma, even if Jørgensen says that we can logically approach the indicative factors of the norms involved in the practical syllogism,

---

[5]See [1, p. 207].

[6]This form of distinction can be found later with the neustic/phrastic dichotomy in [6].

[7]Dependent on the efficacy of the norm. Ross explains: "[...] an imperative I_1 is said to be satisfied when the corresponding indicative sentence S_1, describing the theme of demand, is true, and non-satisfied, when that sentence is false." See [16, p. 37].

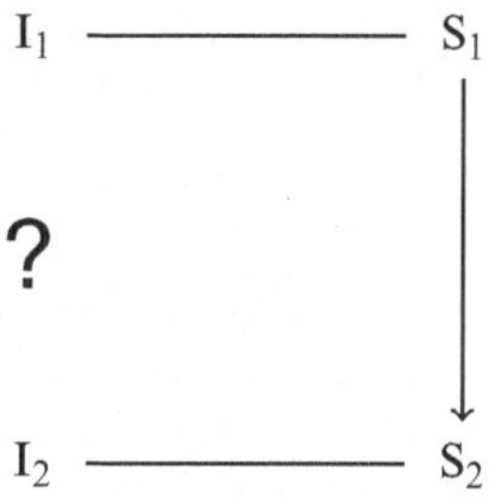

Fig. 9.1 The gap in the practical syllogism

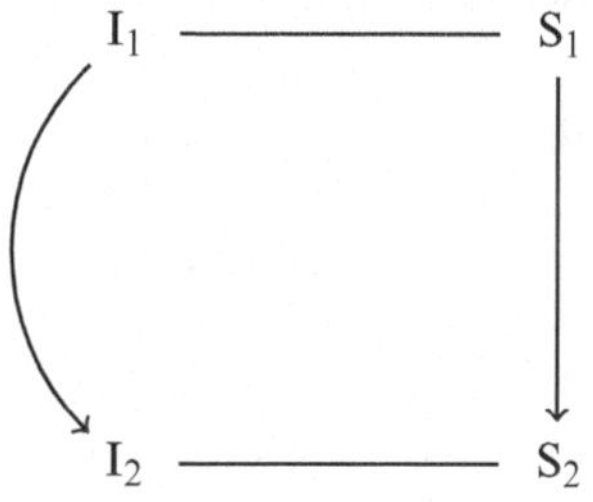

I_1 ——————— S_1

?

I_2 ——————— S_2

Fig. 9.2 The derivation between the two imperatives

I_1 ——————— S_1

I_2 ——————— S_2

nothing allows him to say that we can derive a new imperative from the first one stated in the major premise, even if it is made indirectly. Therefore the dilemma remains unsolved.

To solve Jørgensen's problem, Ross suggests several approaches. First, he suggest an alternative for the apofantic thesis that logic only deals with true/false statements, by saying that the values true/false can be substituted by validity/invalidity, which now should be seen as the logical value of the norms but in a subjective way.[8] The validity is defined by "the presence of a state of mind in the person, that determines this validity", and depends on the will of the imperant. The validity turns into a psychological, not a semantic concept.

Ross starts his paper by stating the problem as being:

> [...] to elucidate whether sentences which are not descriptive, but which express a demand, a wish, or the like, may be made objects of logical treatment in the same or a similar manner as the indicative sentences.[9]

Ross proposes a wider delimitation of the logical domain, by suggesting the substitution of the logical values of truth and falsity by validity or invalidity. So, to fill up Jørgensen's gap in deriving I_2 from I_1 (see Fig. 9.2), Ross declares the following:

> It may then be laid down: if there be any sense in ascribing objective validity and invalidity to imperatives or to a certain group of imperatives, then it is possible to interpret the logical

---

[8]We will analyse this misconception concerning the validity as a value of the imperative or norm later on p. 195.

[9]See [16, p. 31].

deductive system as being applicable to those imperatives. The logical deduction of I_2 from I_1 then means that I_2 has objective validity in case I_1 has objective validity.[10]

But how can we verify the validity/invalidity?

Ross answers that the process of objective legitimation ignores the person and focuses only on the impersonal norm, and this will finally lead to religious morals or to natural law kind of doctrines (imperatives without *imperator*).

Then Ross considers a second possibility, saying that the means of verification of the validity lies on its satisfaction, so that we would have a perfect parallel between an imperative and its satisfaction. In this way, from the imperative "Close the door" we may have the correspondent sentence "The door is closed". Continuing the parallel, we must be capable of having also, for example, the negation of the sentence, i.e. "The door is not closed", from which we should attain the correspondent imperative "Don't close the door". So, in the inference, when we have "Close the door" satisfied, we must have also "The door is closed". Therefore, since the negation is false, its correspondent "Don't close the door" is not satisfied. This example shows the complete lack of similarity between the prescriptive field and the logical field. Moreover, similarly to the negation, a disjunction could be added to the imperative, as in the famous example of "Slip the letter into the letter-box" to which we add the disjunction as: "Slip the letter into the letter-box or burn it". Even if the adding of the disjunction shows no problem concerning the truth values of the correspondent sentences, it is not evident, and even not plausible, that such a disjunction might be relevant in the normative domain, concerning the imperative form (the obligation of sending the letter doesn't imply the obligation of sending it and burning it—according to the inclusive disjunction—because the obligation of sending it is not satisfied if we burn the letter). These examples show precisely how deontic logic cannot grasp the relations between imperatives, and the analogy with the satisfaction is clearly not a good alternative.

Ross then finally proposes the third solution to the problem. The logical aspect of the imperatives in the practical inference concerns not the satisfaction or the verification, but the "subjective validity" of those imperatives. In this case, says Ross:

> An imperative I_1 is said to be valid when a certain, further defined psychological state is present in a certain person, and to be non-valid when no such state is present.[11]

The condition here is that the double validity/invalidity is defined by the truth/falsity of the corresponding indicative sentences, that's to say, the logic directly applied in the true/false sentences would be indirectly applied in the valid/non-valid imperatives when in presence of a determined psychological state. This new element would prevent us from the problems concerning the previous examples such as "Slip the letter in the letter-box or burn it!". The "psychological state" which Ross invokes would serve as a justification or a motivation to the imperant's will, avoiding

---

[10]Ibid., p. 35.

[11]Ibid., p. 38.

trivialities. But this approach is doomed to refer always to a specific case (or a specific trial, for instance) where a norm or an imperative was applied regarding a specific case. Again, the normative relations are put aside, because it is the fact of the creation of the norm that is being asserted in a proposition. As we will see next, in Kelsen's approach, these kind of propositions compose the science of law, i.e. propositions about the validity of the norm, for instance "There was enacted a valid norm saying that 'Murder is punished by imprisonment'", or "It is true that the norm 'X' was applied by Judge Z in case Y". Again, this can be logically treated (in the same way that legal science can be logically treated as any other science), but this does not take into account any relations between norms.

So, according to the transformations between imperative and indicative factors allowed by the rules of satisfaction, the validity of the first norm would be then preserved in the conclusion. But, since nothing guarantees that the person enacting the first order must also enact or want the specific order in the conclusion (in other words, there is no logical—nor legal—necessity between the two norms),[12] a psychological element is required. With this, Ross concludes that:

> Imperatives can be constituent parts of genuine logical inferences, but if so, it is simply a question of a "translation" of logical inferences concerning indicative sentences about the psychological facts which define the "validity" of an imperative.[13]

So, Ross concludes that Jørgensen's Dilemma is the result of a pseudo-logic puzzle. Even if Ross' approach is not restricted to the legal domain, but rather to the prescriptive domain in general, the problem of the legitimacy of the created norm remains the same. And the author recognizes that he does not "solve" the puzzle, but rather shows that it is a pseudo-problem or, more precisely, the case of the use of a pseudo-logic. He says that we simply consider as being a true statement the fact that a norm has been applied by a judge, and obeyed by a person, and the valid imperatives would be transformed in verifiable sentences such as "The judge X has enacted the norm 'Y'", or "This person has accepted the norm 'X'", which does not concern at all the prescriptive level anymore. It no longer represents a practical syllogism, because the norms are being simply described as having or having not taken place. That's why Ross calls it the application of a pseudo-logic.

This discussion deals with many other situations of norm-creation, when the norm is enacted not by the judge or the legislator, but also by any other "authority", such as a father facing his son, or a teacher facing his/her students, or even cases of decision-making which could be reconstructed via a practical syllogism. In the end, all these cases reflect a complete independence of the norm pretending to be the conclusion of the syllogism with respect to the norm admitted as a first premise. This correspondence has no effect on the preservation of the validity from the general norm in the premise to the specific norm in the conclusion. Or, in another words, the

---

[12]The point here is that the specific norm in the conclusion depends on the act of will of the person enacting such a norm, and this process is not achieved through a syllogism. The judge can enact the particular norm without making use of the practical syllogism.

[13]Ibid., p. 45.

imperative in the conclusion is not legitimated or binding *because of* the departing imperative. And Ross himself recognizes that that's where the dilemma emerges: as the result of a psychological mistreatment of the norms.

9.2.2 Kelsen's Battle Against the Dilemma

The works of Hans Kelsen are marked by the constant and close relation to philosophical questions, such as those involving normative justification. It is even a common saying in the juridical area that Kelsen was the most philosophical among the jurists and the most juridical among the philosophers of his time. So, it is not surprising the fact that he devoted a great deal of his attention to the problems concerning practical reasoning, namely Jørgensen's Dilemma.

In the next sections we will analyse how Kelsen treated the previous answers to this problem, and what was his particular solution to it. We must note that Kelsen's approach is marked by his principle concerning the existence of a methodological abyss between the domains of the "Ought" (the normative domain, formed by valid norms, accessed by acts of will—the domain of law) and the "Is" (the descriptive domain, formed by true/false propositions and descriptions, accessed by acts of thought—the domain of science). This conception is essential to understand how, similarly to Ross, Kelsen manages to show that Jørgensen's paradox is an illusion.

9.2.2.1 Correcting Jørgensen

To begin with Kelsen's treatment of Jørgensen's Dilemma, we must at first understand what this author means by the term "modally indifferent substratum".

When considering the differences between the norms ("ought" statements) and descriptions of norms ("is" statements), Kelsen insists that the two cannot be reducible or even comparable. One is the law, and the other is the science that describes the law as its object. Kelsen says:

> 'Is' and 'Ought' are purely formal concepts, two forms or modes which can assume any content whatsoever, but which must assume some content in order to be significant. It is something which is, and it is something which ought to be. But no specific content follows from the form.[14]

So, if we consider the following sentences:

1. "Jérôme C. ought to pay the taxes." and
2. "Jérôme C. pays/had paid his taxes."

---

[14]See [8, p. 58].

We have one and only one modally indifferent substratum, which is "To pay the taxes", under two different modes, namely, the "ought" in 1. and the "is" in 2. Their content is identical, but the sentences themselves are not comparable, correspondent or deductible one from another.

So, concerning Jørgensen's solution to the dilemma, if we stick to Kelsen's conception of the modally indifferent substratum, we must notice that, in fact, the modally indifferent substratum is neither true nor false in itself. It can be the content of a true/false sentence, in the same way that it can only be the content of a valid norm. But there is no correspondence between a true/false statement and a valid norm; they only share the same modally indifferent substratum. So, when Jørgensen says that the answer to the dilemma is only a matter of translation, actually there is no correspondence between an imperative and an indicative sentence: the fact is that we are dealing with the same modally indifferent substratum, presented under two different modes. To sum up, the point of interest is that the modally indifferent substratum is outside the bounds of any logical approach.

9.2.2.2 Correcting Ross

As we have seen, Ross's approach takes as a point of departure the fact that the validity is the value of the norm (or of the imperative) in the same way that truth/falsity are the possible values of the indicative sentence. So, even if Ross's conclusion is that the dilemma is not legitimate since some pseudo-logic is being applied to the imperatives, all his argumentation lies in the presupposition that, when considering practical inferences, the validity is a value of the imperative in the same way that the truth is a value of the indicative sentence. A nowadays critic will resume that:

> Ross directly attacks the idea that the validity, a specific quality of the prescriptive propositions, would be equivalent to the truth, a quality of the indicative propositions.[15]

So, even if Ross's solution is closer to Kelsen extremism in denying the possibility of logical treatment of the norms, Ross takes the wrong way to arrive at this same conclusion.[16]

Indeed, Kelsen will perfectly agree that they are not equivalent at all, but, more than that, Kelsen would also say that the validity is not even a value of the norm. Moreover, still about the differences between norms and sentences, we must note that, when considering this cornerstone of Kelsen's legal theory, we immediately recognize that, actually, the validity can never be predicated from a norm in the

---

[15]The original, in French: "Ross attaque directement l'idée selon laquelle la validité, qualité spécifique des propositions prescriptives, serait équivalente à la vérité, qualité des propositions indicatives." See [2, p. 34].

[16]As we remember, Ross says that the logical treatment in a practical inference is an illusion, since the imperatives are actually treated as indicative sentences (because the evaluation concerns the satisfaction of the norm, which is a verifiable fact).

same way that the truth is a property of a sentence. The validity is not a property of the norm: it is the very fact of the norm's pertinence to a legal system. The norm can only exist, it can only be considered as long as a valid norm. Moreover, while considering positive systems, this is the case for any normative system, including positive morals, and not only positive law.

Moreover the main difference between a sentence (*Sätze*) and a norm (*Sollsätze*) is that the first is the expression of an act of thought, and can be evaluated as true or false, and the second is the sense of an act of will, and can only be valid. This main distinction allows Kelsen to give a more precise treatment than Ross with respect to the impossibility of deriving a binding imperative from a previous one: the "second" is the result of nothing but an act of will, whose sense is completely independent from the act of will giving place to the first norm or imperative. This leads Kelsen to explain that, even if the two imperatives display the same content (the same modally indifferent substratum), the acts of will must be two, neatly separated and independent from each other.

Before moving to the next section, it is important to note that it is clear from Ross's writings that he does not pretend to approach the practical syllogism from a legal perspective, such as Kelsen does, but rather from the perspective of our daily situations of decision-making, or when confronted to imperatives in general. The problem, in this case, would not be the issue of the legal objective validity of the norm "obtained" in the conclusion of the practical syllogism, but rather its legitimacy, i.e. the fact of its bindingness given the first premise. The same question is posed in both approaches, namely, what renders this imperative in the conclusion obligatory or binding in relation to the first norm presented as a premise? Why do I, once having accepted the first imperative, must also obey the second?

9.2.2.3 Kelsen's Final Solution

So, after having seen Kelsen's position in front of the two answers given to the dilemma, how would Kelsen himself solve this puzzle? The answer lies in the difference between an act of thought and an act of will.

To Kelsen, every norm must have content, the "modally indifferent substratum", which is not the indicative factor as Jørgensen believed, but the behaviour wanted by the *imperator*. So, in one way, the person enacting the norm knows the content of the norm, his will is directed to some behaviour, indicated by the norm. On the other way, the addressee of the norm must also understand what is he supposed to do, how is he supposed to behave, and he does that by observing what is the content of the norm. Let's see how it works in an example. Consider the following picture.

Every citizen should pay the taxes. This applies also to, let us say, the French citizen Jérôme C. Once a year he receives a letter saying how much he has to pay to the government, as income taxes. The legal norm behind the letter, put in a simpler way, says that "Every citizen must pay his/her taxes", even if this very prescriptive element is not explicitly formulated in the text addressed to Jérôme C. The norm-positing subject can justify his demanding for the money by enumerating the benefits to beseized by the citizens, like the improvement of social and public services, the reduction of budget deficits etc. In Kelsen's theory, those are to be seen as acts of thinking preceding the norm, and they have nothing to do with the will to have the taxes paid. They have nothing to do with the modally indifferent substratum, which is, in this case, "To pay the taxes".

But let us see how Jérôme C. could be confronted to the imperative directed at him, namely "Every citizen must pay his/her taxes". Once confronted to this general norm (the first premise in the misleading "practical syllogism"), Jérôme C. can have acts of thought which allow him to understand what is he supposed to do, given his specific situation. He can think about things like "If a pay my taxes, my equals will benefit from the improvements in education, security, health care and so on. . . ". Also, he can consider that "If I don't pay the taxes, and if I figure out how to keep it secret, I can benefit myself from my own money. . . ". Outside the motivation frame, he must first of all understand that he is obliged to pay the taxes: the addressee of the norm must know what he's supposed to do. So, he also has acts of thinking which allow him to comprehend things like "it's me and no other person who has to pay these taxes", "I'm supposed to pay what it's said in the letter, not more nor less", and those are true/false statements, those are the sense of acts of thought. They don't belong to the norm in question; they are *about* the norm enacted. After understanding all that, Jérôme C. can accept that, once "Every citizen must pay his/her taxes" is valid (the sense of someone's act of will), and after the arriving of the letter from the tax collection, he must pay his taxes, i.e. he must perform himself the general norm which was, by the letters sent to him, applied to his case, understanding that "Jérôme C. must pay his taxes"—otherwise he will be in trouble. This concerns a fundamental division between thinking about possible norms, and wanting a norm, i.e. enacting it.

Concerning Jørgensen's approach, Kelsen says that the indicative factor pointed by this author has in fact a complete different meaning than the imperative factor. The indicative factor (the sentence) is the sense of an act of thought; the imperative factor (the command) is the sense of an act of will, and the two are completely independent from each other.

This is how Kelsen solves the dilemma. The misleading "practical syllogism" actually works in our daily lives because we are in fact reasoning about the norms involved in the inference, through acts of thought. It's natural that we try to understand what the norm is demanding, in order to know how to behave. The minor premise which poses a fact is actually the context in which we reason, and the major premise (indicating a norm in the practical syllogism) poses the norm that we must understand, comprehend, which presents the behaviour wanted. If I understand the first premise in the context of the second premise, I'm able to know which behaviour I'm supposed to have. All this procedure does not touch the normative level, because no act of will has yet being considered or enacted. It's all a question of knowing what is at stake, through mental processes. Also, it's perfectly possible to understand the content of a general norm and not want to observe the behaviour posed as obligatory. The particular norm posed as a conclusion in the practical syllogism will only be a valid norm after an objective act of will.

9.2.3 Any Objections to This? The Case of von Wright's Deontic Logic

When considering von Wright's first attempts to construct a logic of/for norms, we are confronted with definitions of the following type:

> If an act is not permitted, it is called forbidden. For instance: Theft is not permitted, hence it is forbidden. We are *not allowed* to steal, hence we *must not* steal.[17]

What von Wright tries to do is to use logical equivalence to established the three deontic modes (obligatory, permitted and forbidden). In fact, given a primitive modality P (standing for "it's permitted that...") and an arbitrary proposition φ, it is easy to define 1. a modality *forbidden* as $\neg P\varphi$, i.e. it is not permitted that φ, so φ is forbidden, and 2. a modality *obligatory* as $\neg P\neg\varphi$, i.e. it is not permitted that not φ, so φ is obligatory. This is exactly what von Wright does in his example: "forbidden" is equivalent to "not permitted". But Kelsen says that the only normative function is the mandatory function, norms have no permissive function. When von Wright says that "We ought to do that which we are not allowed not to do", he defines the obligatority of an act by the negation of the permission to not doing this act.[18] Norms start from the opposite, when something is commanded (mandatory normative function), the fact of abstention is qualified as not permitted. Everything we need is only the mandatory function, everything else is accessory.

Let's see what Kelsen himself says about it:

> Wright says in this connection: "If the negation of an act is forbidden, the act itself is called obligatory. For instance: it is forbidden to disobey the law, hence it is obligatory to obey the law" (1951:3; 1957:61). Since the normative function being considered is that of commanding, and since to forbid is to command an omission, things are being stood on their head when the being-commanded of an act is presented as the being-forbidden of its omission. "It is commanded to obey the law": if we want to make use of the concept of forbidding, we can express the same idea by saying "It is forbidden not to obey the law". But things are back to front if we say "It is forbidden not to obey the law, and *hence* it is commanded to obey the law."[19]

Kelsen is right in saying that it seems strange to define what is obligatory by the negation of what is not permitted. But from a strictly logical point of view, if what

---

[17]See [18, p. 3].

[18]See [18].

[19]See [8, p. 322].

is permitted can be defined by what is not obligatory, conversely, what is obligatory can be defined by what is not permitted to not doing. It seems to be a tension between logical and legal interests. Moreover, in the introduction of von Wright's book,[20] which includes a re-edition of "Deontic Logic", the author says that, about this last essay, a further refinement is needed in the treatment of this question:

> Another application is to the logical study of the norms (normative discourse). This latter study is important to ethics and the philosophy of law. But it must be pursued with much more refinement than in my first paper (here republished) on deontic logic. Philosophically, I find this paper very unsatisfactory. For one thing, because it treats of norms as a kind of proposition which may be true or false. This, I think, is a mistake. Deontic logic gets part of its philosophic significance from the fact that norms and valuations, though removed from the realm of truth, yet are subject to logical law.[21]

This confession shows that von Wright actually commits the same mistake as his predecessors, Jørgensen and Ross. When he says that, even being "removed from the realm of truth", norm are still submitted to logic laws, he is actually stipulating some kind of strange correspondence between truth and legal validity.

9.3 Dialogs About Kelsen's Solution

Kelsen's approach to Jørgensen's dilemma gives us some insights about how to deal with the justification of legal decisions in the legal argumentation. A legal decision involves a general norm that is applied by the Judge in relation to a specific case (fact) in order to "create" a specific norm for that case. We all know now that the creation of a legal norm depends on an objective act of will coming from an authorized person, and no logical element interferes in this procedure. If we are about to deal with the normative reasoning (or the normative argumentation), the fact of trying to compare and correspond (legal) norms to (logical) propositions will forcedly lead to error. But with the notion of modally indifferent substratum, Kelsen shows that it is fairly possible to reason about norms, to evaluate our behaviour in relation to a norm, without disrespecting the limits between these two domains. This is precisely what we aim to do: to show how is it possible to logically discuss and reason about norms, and still maintain the dichotomy between law and its science.

If we were to draw a scheme concerning the modally indifferent substrate (Fig. 9.3), we must insist on the fact that this notion allows a double treatment of the normative content (paying-taxes) in terms of a (possibly) valid norm in the same way that it might assume a form of a true/false sentence. In the first case the dialogical approach to logic will deal with the normative construction in terms of a *Procedural justification*, while in the second case the treatment will be made

---

[20]See [19].

[21]See [19, p. vii].

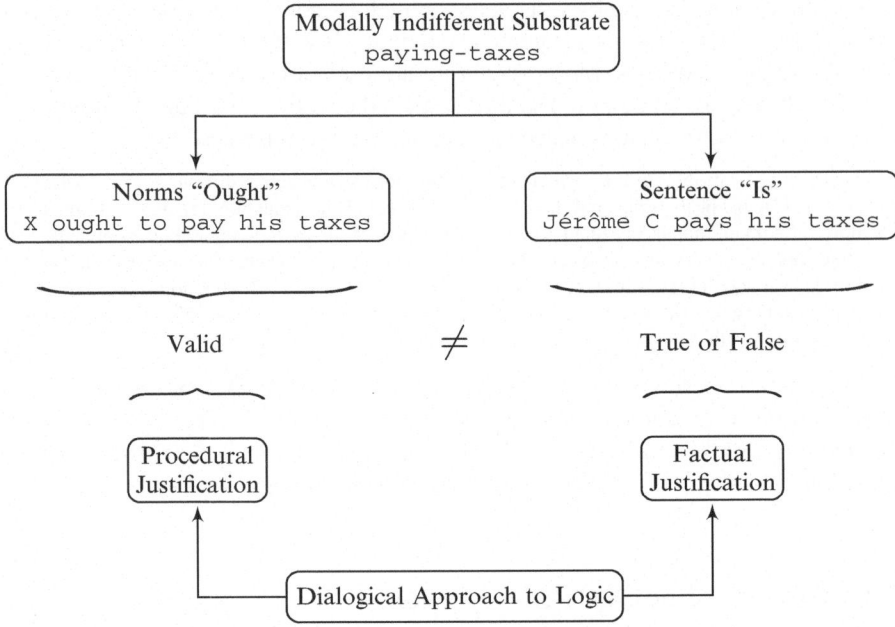

Fig. 9.3 Modally indifferent substrate and dialogical approach to logic

in terms of a *Factual justification.*[22] The main point is that both the legal and the logical treatment respect and maintain the distinction between the two forms which can assume the same content.

9.3.1 Law and Legal Science in Dialogs

Regarding the process of normative creation, Jørgensen and Ross insist in considering the legal reasoning by forcedly comparing, corresponding, translating sentences into norms and vice-versa. In the end, this results in disturbing structures such as the practical syllogism. But once we have the legal material (norms used by the Judge and the facts), if we use the (*dia*)logical approach to logic it becomes not only possible to deal with these two separated domains but also to justify the legal procedures by means of an argumentative process (as it is made in legal science). This purpose does not aim for a logical treatment of norms by means of normative creation via logic, but rather for the logical analysis of the already created norms, i.e. of the sentences about their validity.

---

[22]See Sect. 9.3.1.2.

9.3.1.1   The Dialogical Approach to Logic

The dialogical approach to logic was first presented in the end of 1950s by Paul
Lorenzen, and further developed by Kuno Lorenz.[23] During the 1990s, this kind of
dialog was adapted and applied to different classical and non-classical logics.[24]

The main idea of this approach is inspired by Wittgenstein's maxim of the
"meaning as use". Following this idea, one can define a logical constant by its use
made by two protagonists through an argumentative procedure. According to this
argumentative procedure, a player proposes a thesis which he will have to defend
against the other player, the latter being opposed to the thesis by trying to conceive
a counter-argument. These two protagonists use the same rules, with one exception:
the formal restriction.[25] This rule forbids Proponent to utter an atomic formula (i.e.
a formula which does not contain logical connectives) if Opponent has not already
uttered it. This restriction, in spite of its asymmetry, assures the logical truth, if
Proponent manages to defend himself from anyone of his adversary's attacks.[26] This
rule guarantees that Proponent's use of an atomic formula depends on a justification
procedure, that is, the use of this atomic formula is independent of the notion of truth
or falsity. Consequently, if Proponent wins,[27] the thesis is a logical truth with respect
to the dialogical system used. Moreover, during a dialog, a player can only attack the
logical structure of his adversary's discourse, never the content of the discourse, nor
even the adversary himself. If in this form of dialog the players commit themselves
to defend their arguments announced during the play, those commitments are not
about the argument's content, but about their logical form. In other words, Opponent
and Proponent are only committed to defend the logical structure of the propositions
they use.

From a technical point of view, dialogic is not a specific logic, but rather a
conceptual framework providing the study of different logics under an argumentative process. Dialogical language is defined from a determined logic where the
letters **O** and **P**, standing for Opponent and Proponent, are added as well as the
symbols "?" and "!". These symbols are required in the formulation of the rules of
use of the logical constants: the *particle rules*. Another set of rules determine the
conditions under which the particle rules have been or can be used. These are called
the *structural rules*.

---

[23]See [9].

[24]This approach allows the combination and analysis of different logics in one and the same frame.
See [13] for examples or [7] for a general view of these different developments.

[25]The overall set of used rules settles the dialog system. See Appendix p. 219 for a presentation of
the standard rules of the dialogical approach to logic.

[26]Through the notion of winning strategy.

[27]See Winning Rule **SR-3**, Appendix p. 219, §Structural Rules for the definition of the principle
regulating victory.

Modal Dialogic

Modal Dialog results from an original idea coming from the researches of S. Rahman and H. Rückert.[28] This kind of dialog contextualizes the notion of proposition. If the standard dialog allows us to define the meaning of the logical constants through their use in an argumentative process, standard modal dialog allows us to associate a contextual dimension to this notion of use. The meaning of a logical constant is directly dependent on its contextual use. Not only the meaning is defined by its use, but this use is also dependent on a context. The context is then a constitutive part of the meaning of the constant. Moreover, the pragmatic character of the dialogic semantics displays different interpretations of the modality simply by changing the associated structural rule. To our purposes, the modal operator receives a deontic interpretation, that is, we interpret the modal operator $O\varphi$ as "it is obligatory that φ" or "φ ought to be the case".

9.3.1.2   Dialog, Validity, Truth and Justification

In this section we will present the formal details of our dialogs. But before that, we must explore both the advantages and inconvenient aspects of the dialogical approach concerning our problem, because if the dialogical frame allows an argumentative approach to logic, this frame is more concerned with the formal truth of a formula than its truth itself. This the reason why we need some modifications of the standard dialogical framework. The next paragraphs lay the groundwork for our conceptual modifications.

Material Dialog and Justification

The dialogical approach to logic deals with the logical validity of a formula and not its truth value. But legal science, the discourse about the law, is evaluated in terms of truth rather than logical or legal validity. Besides, different approaches of the dialogic allow to characterise the truth value of a formula in *material dialog*.[29] First of all, we need to consider initial concessions, i.e. some hypothesis over which the dialog will be constructed. Those hypotheses can be used by Proponent to justify his utterance. Thus, while in the dialogical approach we only consider Proponent's justification via the procedural use of atomic formulas, that is, we consider his saying: "I use this formula in the same context and in the same way you did", with material dialogs he can justify a proposition saying: "I use this proposition because it is an initial concession/premisse". This tends to weaken the formal restriction:

---

[28]See [14].

[29]See [15].

Proponent is no longer restricted by this rule, he can use the initial concessions to utter and/or justify his use of atomic propositions although Opponent has not already utter them before.

Dialog About Truth

From a logical point of view, material dialog deviates from the notion of winning strategy—which is, in dialogic, the counterpart notion of validity. With the material dialogs, a winning strategy for a given thesis becomes relative to a set of premises. It's only with regard to the set of premises that the winning strategy exists. So, it's possible to show that the thesis holds with respect to the premises given before the dialog, and this fits perfectly with the kelsenian purpose saying that the legal theory is a true/false discourse raised from law. Thus, even if this kind of dialog keeps us apart from the notion of logical validity, we stick to Hans Kelsen conception that the legal science only deals with truth.

Validity and Truth?

According to Hans Kelsen, legal science—the discourse about the law—has to be supported by the law itself, i.e. a norm or a set of norms.[30] In the same way, the two players of the dialog discoursing about the law (about the normative creation or the legal decisions) have to support their arguments over this normative set and over what the Judge qualifies as facts according to a determined case.[31]

In a given logical system a rule is not true nor false, but valid. A rule is valid inasmuch as it holds independently of its application. However, the particular application of this rule can be true or false, because it requires certain conditions to be fulfilled. If these conditions are not fulfilled, the rule cannot be applied, but this does not mean that the rule is invalid or false. The rule remains valid even when it's not the case that it has been or can be applied in the overmentioned circumstances.[32] In fact, the distinction between logical rule and particular proposition is analogous to the distinction between valid legal norm and contingent legal proposition. So, this distinction can be maintained and reconstructed if we distinguish two sets of premises: a set of general valid rules and a set of contingent propositions, each one of those representing, respectively, the set of norms and the set of facts.

---

[30]The "Civil Code" or the "Constitution". See [8, pp. 61–62 and 81–82].

[31]Consequently, the corresponding dialog must beforehand suppose as admitted both the set of norms and a set of facts. This fits perfectly with the notion of initial concessions just mentioned.

[32]We will come back later to the discussion over validity and truth, with the justification notion.

Validity and Truth: A Difference Concerning Justification

Previously, we have put in evidence a crucial aspect, namely the incidence that a set of premises has in respect to the question of the justification of the atomic proposition. Standard dialogic limits itself to a procedural approach of the justification. It means that Proponent's use of an atomic proposition depends on a previous use of this proposition by Opponent. Implicitly, Proponent can justify its use by copying the procedure that Opponent could possibly follow to justify this proposition.[33] Material dialogs allow a justification for the use of an atomic proposition, but not due to its previous use by Opponent. Rather, the justification can be made because the proposition belongs to the set of initial concessions. Because of this authorization for this kind of justification, material dialogs are closer to the notion of truth and farther from the notion of validity. However, nothing prevents Proponent to use the procedural form of justification to justify an atomic proposition which do not belong to the set of initial concessions. In fact, it is possible to remove the formal restriction and to distinguish two distinct forms of justification: the procedural justification and the propositional justification. On one hand, the propositional justification allows to produce a justification with respect to truth, and on the other hand, the procedural justification has the same effects of the formal restriction and consequently preserves the (logical) validity level. The combination of these two kinds of justification in a same frame allows us to deal with the truth of a particular discourse and with the validity of a rule, without reducing the first to the latter and vice versa.

9.3.2   The (Dia)logical Tools

We have chosen to benefit from the dialogical frame in order to maintain the distinction between law and legal science. But before developing these formal details, we have to specify first the dialogical modal frame under which our reflections are elaborated.

9.3.2.1   Preliminary Notions

With respect to standard propositional dialogic (presented in the Appendix), we have to add two operators in order to obtain a dynamic modal dialogic. The first operator corresponds to the modal operator O (previously mentioned in the Sect. 9.3.1.1,

---

[33]Thanks to the justification rule that we introduce, this procedure becomes explicit. See p. 208.

§Modal Dialogic). The second operator is the dynamic operator $[\varphi]\psi$.[34] We interpret this operator as: "if φ is proved, then ψ holds".[35] Whereas the introduction of a modal operator in a dialogical language demands the contextualization of the players' utterances, the dynamic operator prefixes the chosen contexts through a list $\mathscr{A}$.

The Modal Operator O

A contextual point is a positive integer labelling a formula.[36] The particle rule of the modal operator, presented in Table 9.4, allows to determine who between the challenger or the defendant will support the burden of the choice of the contextual point. With respect to our modal operator, the burden of the choice amounts to the challenger. Intuitively, this rule captures the following exchange: "if the player **X** utters, in the contextual point i, it is obligatory that φ, the player **Y** can choose a contextual point j in which **X** must defend that φ holds." The structural rule **SR-O**, displayed in Table 9.5, defines the conditions for the choice of the contextual point.

The Dynamic Operator $[\varphi]\psi$

In the particle rule as well as for the structural rule of the operator **O**, the contextual point i is prefixed by the list $\mathscr{A}$. The necessity of this list appears in the particle rule of the dynamic operator (Table 9.2) and its associated structural rule (Table 9.3).[37] The list $\mathscr{A}$ comes to enrich the notion of contextual point. Strictly speaking, it is not a new contextual point, it remains the same but prefixed by a list of ordered set of formulas. The list allows to keep track of the formulas coming from the dynamic operator.[38]

---

[34]Commonly known as the Public Announcement operator in Dynamic Epistemic Logic (DEL). See [4] for a DEL overview and [11, chaps. 6–8], for more details about the juridical use of this dynamic operator.

[35]The dynamic operator entails a conditional form, but this conditional form is far from the well-known material conditional. The consequent requires that the antecedent is true. If the antecedent is not true, it cannot be announced and the consequent cannot be evaluated in the submodel where the antecedent was true before its announcement. See [4, chap. 4] 4 for more details.

[36]A contextual point is not an atomic formula but it receives the same restriction, **P** only being capable of re-use those introduced by **O**.

[37]These rules are originally introduced in [10]; soundness and completeness proof of them is given in [12].

[38]See [11, chap. 5].

Table 9.2 Dynamic operator – **PR-DO**

Burden and/or object of choice	X-Utterance	Y-Challenge	X-Defence
$[\varphi]\,\psi$, the defender has the choice	$\mathscr{A}\|i : [\varphi]\psi$	$\mathscr{A}\|i : ?_{[\,]}$	$\mathscr{A}\|i : \neg\varphi$ or $\mathscr{A} \bullet \varphi\|i : \psi$

Table 9.3 Structural rule **SR-A**

For any move $\langle X-\varphi_1 \ldots \varphi_n\|i : e\rangle$, player **Y** can compel **X** to utter the last element φ_n of the list $\mathscr{A}$:

SR-A

- In the contextual point $i : \langle Y - \varphi_1 \ldots \varphi_{n-1}\|i : !_{(\varphi_n)}\rangle$ or
- In the contextual point $j : \langle Y - \varphi_1 \ldots \varphi_{n-1}\|j : !_{(\varphi_n)}\rangle$ if $e = ?_j$

Because of the "if...then" form of this dynamic operator, the burden of the choice is supported by the defender. When the player **Y** challenges the evidence φ, **X** can reply "φ is not the case" or "because of the evidence φ, ψ is the case".[39]

The structural rule **SR-A**, defined in Table 9.3, specifies certain conditions for the use of the list. The general idea of this rule is to ensure that a player who adds a formula in the list $\mathscr{A}$ is able to justify this formula.[40]

The rules presented above generate a dynamic modal dialogical frame. This dialogical frame allows to evaluate the logical truth of modal dynamic formulas.

9.3.2.2   The Set of $\mathscr{F}$ acts and the Problem of Justification

In order to reach the truth level, we have to introduce the sets of premises called $\mathscr{N}$ orms and $\mathscr{F}$ acts[41] in the dialogical frame obtained, and re-found the question of justification of atomic formulas.

Definition 1 (Fact) We take as a fact all data accepted by the Judge. These data are expressed by formulas in the set called $\mathscr{F}$. This set is fixed before the dialog starts, and these formulas can be used in any contextual point of the dialog.

---

[39]The dynamic operator $\langle\varphi\rangle\psi$, such that $\langle\varphi\rangle\psi \overset{\text{def}}{=} \neg[\varphi]\neg\psi$, expresses the same idea but with a conjunctive form. So, in this case the burden of the choice is not supported by the defender but by the challenger. See [10] and [11, chap. 3].

[40]This rule is counterpart of the fact that only true formulas can be used with this dynamic operator—what can sound a bit idealized if we consider evidence. An interesting point would be to consider refutable evidence, but to do this we need to use more sophisticated dynamic operators.

[41]As discussed in Sect. 9.3.1.2, §"Validity and Truth ?".

Burden and/or object of choice	X-Utterance	Y-Challenge	X-Defence
O, the challenger chooses a contextual point i'	$\mathscr{A}\lvert i : \mathbf{O}\varphi$	$\mathscr{A}\lvert i : ?_{i'}$	$\mathscr{A}\lvert i.i' : \varphi$

Fig. 9.4 Modal Operator **O** – **PR-MO**

SR-O   To challenge a move $\langle \mathbf{O} - \mathscr{A}\lvert i...i' : \mathbf{O}\varphi\rangle$, **P** can choose any contextual point i'' already chosen by **O**.

Fig. 9.5 Structural rule **SR-O**

Those facts represent the data accepted by the Judge. For instance, in our example, "Jérôme C. has committed tax evasion" or "Jérôme C. has not committed tax evasion" are facts.

Recall:

the two players can utter atomic formulas because we have eliminated the formal restriction. However, in order to maintain an internal coherence of the dialog, if a player uses an atomic formula, his adversary can compel him to justify it. That's why a justification rule is required.

The justification rule (**PR-J**) allows distinguishing between two different kinds of justification:

1. Propositional justification, and
2. Procedural justification.[42]

This differentiates a justification which is based on the set of facts, from another which is founded in the argumentative process itself. It's possible for the latter kind of justification to not even be based on the facts.[43] A consequence of this is the fact that we need to add the following logical constant in our logical language:

$$\llbracket p \rrbracket \text{ for all atomic propositions,}$$

where $\llbracket p \rrbracket$ is a justification for the atomic proposition p.

---

[42]Even if we present it in two distinct tables (Figs. 9.4 and 9.5), it is in fact one and the same rule authorizing two different defences.

[43]We come back to this distinction in the paragraph "Propositional and Procedural Justification", p. 208.

Table 9.4 Propositional justification rule

Propositional justification	X-Utterance	Y-Challenge	X-Defence
$\mathscr{A}\|i \ : \ p$, the challenger requires a justification for the proposition p	$\mathscr{A}\|i : p$	$\mathscr{A}\|i : !_{[\![p]\!]}$	$\mathscr{A}\|i : [\![p]\!] \in \mathscr{F}$ if $p \in \mathscr{F}$

Table 9.5 Procedural justification rule

Procedural justification	X-Utterance	Y-Challenge	X-Defence
$\mathscr{A}\|i \ : \ p$, the challenger requires a justification for the proposition p	$\mathscr{A}\|i : p$	$\mathscr{A}\|i : !_{[\![p]\!]}$	$\mathscr{A}\|i : [\![p]\!] \in \mathscr{M}_Y$ si $\langle \mathscr{A}\|i : p \rangle \in$ Y-move

1. Propositional Justification.

A propositional justification can only be produced according to the set of facts $\mathscr{F}$, i.e. an atomic proposition can only be propositionally justified if it belongs to the set $\mathscr{F}$. This rule is described in Table 9.4.

The set of facts ($\mathscr{F}$) and the rule for the propositional justification allow establishing a perfect symmetry between Proponent and Opponent with respect to the moves they can make in the dialog. So, the two players have to use strictly the same rules.[44]

2. Procedural Justification.

In order to fully understand the rule of procedural justification, let's suppose that, during the play, player **Y** introduces an atomic proposition and, then, player **X** utters this same atomic proposition. Player **Y** can compel **X** to provide a justification for this atom. In this case, **X** can defend himself by copying **Y**'s (possible) justification. That is possible only if this proposition belongs to **Y**'s moves in the play, what we note as **Y**-moves ($\mathscr{M}_Y$) in the rule displayed in Table 9.5.[45]

The procedural justification consists in justifying the use that a player makes of an atomic proposition by a previous use made by his adversary in the game. Thus, when a player use this kind of justification, he affirms nothing more than "it's legitimate for me to use a proposition in the same conditions as you did".

Propositional and Procedural Justification

While the propositional justification allows a link between an atomic proposition used in the dialog and its truth value fixed in the set of facts, the procedural

---

[44]Normally, the two players do use the same rules, but the formal restriction introduces an asymmetry. Our justification rule allows establishing a strict and complete symmetry.

[45]This rule gives a typical copycat procedure, i.e. a procedure identical to the formal restriction.

justification allows us to use propositions that do not necessarily belong to the set of facts ($\mathscr{F}$). The conditions of use of these propositions are identical for the two players. During the game, if a player introduces a new proposition, he authorizes his adversary to use this proposition under the same conditions and according to the same justification. Concerning the argumentative aspect, the main focus is not over the fact that a proposition belongs to the set $\mathscr{F}$ (consequently, over its truth value), but over the symmetry of the conditions of use of these propositions, i.e. whatever **X** does, **Y** can do the same.

Moreover, this strict symmetry between the two players let us consider the two players as having the same cognitive capacities (they can make the same inferences) and identical argumentative means (there is never imperfect or hidden information). From a strict juridical point of view, it would be difficult to consider an argumentative debate over law where the two protagonists won't have the same argumentative tools.

The difference between these two kinds of justification is illustrated by our example. In Table 9.8, it is the propositional justification which is used by Proponent, while in Table 9.10 he uses the procedural one.[46]

Factual Truth and Formal Truth

The choice with respect to the nature of justification—propositional or procedural—is crucial because it permits to distinguish two levels of truth: factual truth and formal truth. On one hand, the propositional justification is based on the link between the propositions uttered during the dialog and the facts (admitted in $\mathscr{F}$). Thus, this kind of justification allows us to establish a correspondence between the discourse and the facts, which is precisely the definition of truth. On the other hand, the procedural justification doesn't focus on the discourse's truth relatively to the facts. Procedural justification is an internal form of justification, independent from the truth of the facts. It's based on the formal truth of the propositions, on the propositions' structure independently of their relation to the facts. This reveals a difference between the true propositions with respect to the facts, and the justifiable propositions with respect to a given procedure. If an atomic proposition belonging to the set $\mathscr{F}$ is necessarily true and can always be justified, the same doesn't happen with a justified atomic proposition, which is not therefore necessarily true. The procedural justification authorizes a justification independently of the propositional or factual truth.

The notion of justification is then founded over a larger notion than that of truth. It is founded over a player's ability to defend a proposition, so that a player can defend himself without using the truth notion. It's the player's use of the justification rule which finally determines the discourse level.

---

[46]See Sect. 9.3.3.1, p. 213 and following.

9.3.2.3 The $\mathcal{N}$ormative System

Definition 2 (Norm) For "Norm" we take the law used by a Judge and recognized as so by the two protagonists of the dialog. Those norms are formulated by general rules and constitute the set of norms $\mathcal{N}$. This set is fixed before the beginning of the dialog and the general rules can be used in any contextual point of this dialog.[47]

A particular norm is produced from a more general one. This process passes by the Judge, who has a crucial role in the creation of the particular norm. He needs, on the one hand, to be based in the existence of a general norm with a corresponding content and, on the other hand, the particular fact concerning the affair in question. From that, once authorized by law itself, he is able to create a particular norm. According to Kelsen, a legal authority can fill up this role of creation of a particular norm. In the kind of dialogs we deal with, those general norms are reconstructed by the logical rules which determine the set $\mathcal{N}$, i.e. the logical set of general rules should be seen as the counterpart of the legal set of general norms. These norms can be used by the players during the dialog, but under certain conditions, which are specified in the following paragraphs. Before that, we have to mention a particularity concerning the use of the legal order in a dialog.

When a player wants to use one of the general norms of the set $\mathcal{N}$ (Table 9.6), he expresses a *question* about the use of this general norm (which was used or could be used by the Judge). It cannot be a challenge against a previous move because players are not allowed to create particular norms, only the Judge is allowed to do this. So, the rule must be read in the following manner: "from a general norm belonging to the legal order, any player can ask his adversary if the Judge has used or could have use this norm to take his decision over an individual, which is represented in the rule by the choice of the individual constant a."

In the formulation of the rule for the use of $\mathcal{N}$, the predicates A and B are used to, respectively, designate an Act made by an agent x and a Behaviour to be imputed to the same agent x concerning this act. The "Ought" of this behaviour is translated by the operator O presented in the Sect. 9.3.2.1.

Table 9.6 Rule for $\mathcal{N}$

	$\mathcal{N}$	**Y**-Question	**X**-Defence
$(Ax \rightarrow OBx) \in \mathcal{N}$, **Y** asks if the norm can be applied to the agent a	$\mathscr{A}\|i : Ax \rightarrow OBx$	$\mathscr{A}\|i : ?_{a/x}$	$\mathscr{A}\|i : Aa \rightarrow OBa$

---

[47]The set $\mathcal{N}$ cannot contain conflicting norms. This aspect is somewhat idealized, because it is not impossible for the norms to enter in conflict. The question concerning conflicts between norms and their choice is not of our interest by now, but could certainly be the object of future researches.

N ormative System and Its Conditions of Use

If **Y**, in his question about the set *N*, can choose an individual constant, he's not therefore allowed to arbitrarily introduce a new individual constant in the play. The individual constant chosen must be previously given, i.e. at least one occurrence of the individual constant must occur in the play. But the simple occurrence of the constant is not yet sufficient for the player to be able to request the set of norms. The antecedent of the demanded particular norm, containing the individual constant in question, must be a justified element in the list *A*. Thus, the particular norm Aa → **O**Ba can only be obtained during the play if the defence for [Aa]ψ leads the player to add Aa to the list *A*, that is, if there is an evidence that a satisfies the antecedent of the norm.

In fact, every justified proposition added to the list is called an evidence. But, since a proposition in the list *A* can be justified simply by its belonging to the set of facts *F* or, for those which do not belong, by the procedural justification, two kinds of evidence need to be distinguished:

1. *A* ∩ *F* ; and
2. *A* \ *F*.

The set *A* ∩ *F* describes the set of propositions belonging to the list *A* and to the set of facts, while *A* \ *F* delimits the set of propositions belonging to the list *A* minus the propositions belonging to the set of facts. The justification of a proposition of the list by the set of facts (1) ensures that the demanded particular legal norm is created by the Judge with respect to the facts established and accepted by both players. So, the Judge can create the particular norm because it is established that the act committed by a is a fact. If a general norm imputes a behaviour B according to an act A, and if it's observed that, in the facts, the individual a has committed the act A, the Judge can create the particular norm imputing the behaviour B for individual a. However, for the propositions which do not belong to the set of facts (2), the justification can only be procedural. The subsequent evidence is not based on a veridical discourse (held by the notion of truth), but on the ability to defend by using the adversary's arguments. This warrants the validity of the Judge's created norm, independently of his having recourse to the factual truth.

What Is and What Could Have Been

The propositions in the list *A* depend on the players' ability to justify them. If we consider that the set *F* attributes a truth value to propositions belonging to it, propositions added to the list and justified using this set (propositional justification) describes *what is*, whereas propositions added to the list using the procedural justification represent *what could have been* or *could be* the case, but which is not the case with respect to the facts. From this, we can go on and distinguish two levels of discussion between Opponent and Proponent:

1. For all propositions verifying $\mathscr{A} \cap \mathscr{F}$, the players discuss about the particular norms created by the Judge with the established facts—the truth concerning the application of the particular norm with respect to the established facts.
2. For all propositions verifying $\mathscr{A} \setminus \mathscr{F}$, the players discuss about the particular norm that the Judge could have created or could create if the antecedent of the norm were/becomes an established fact—the validity of the general norm via the possible particularization independently of the facts.

From a strictly grammatical point of view, in the second level of the discussion, the question posed about the particular norm created by the Judge changes its mode. The first level of discourse corresponds to the indicative tense: there is a norm created by the Judge. The second level of discourse is performed under the past or present conditional tense: he could or could have created a norm such and such...This changing of mode is interesting because it puts in evidence the conditional aspect of the normative creation regarding the existence of the fact. It is true that the general norm is formulated under a conditional proposition, but its use is itself conditioned by the existence of a fact corresponding to the normative antecedent. If the antecedent belongs to the set of facts, the grammatical mode is the indicative tense: Opponent and Proponent start the debate from the existence of a particular norm whose creation was linked to a fact. But, if the antecedent belongs to the list of announcements, but not to the set of facts, then the grammatical mode is the conditional tense. Opponent and Proponent consider the particular norm that could be or could have been created if the existence of the fact was or could someday be recognized. The procedure to follow is identical being the fact an existing fact or not: the difference lies in the level of the mode and the justification. To the existing facts, it is the propositional justification which takes place (or the procedural justification, if the propositional justification was already used). To the non-existing facts, only the procedural justification can take place. So, the focus is not over the existence of the fact, but over the players' ability to justify their use of those facts. The players do "as if" the fact existed, even if it is not actually the case. In that case the fact is only used as an element inside a procedure.

The logical formalization which we introduced allows Opponent and Proponent to discourse about the particular norms created by the Judge, according to the facts and to the norms that could or could have been created if the existence of the fact was proven. In order to emphasize this distinction, we distinguish in the list what is justified by the set of facts from what can only be justified by a procedural means. For every proposition added to the list without being propositional justified (i.e. all proposition $\varphi \in \{\mathscr{A} \setminus \mathscr{F}\}$), we add $*$, such that φ^* means φ can only be procedurally justified. Thus, in a play, the propositions $\varphi_1, \ldots \varphi_n$ in the list $\mathscr{A}$ have a truth value, while the propositions $\varphi_1^*, \ldots \varphi_n^*$ have a value determined in terms of their capacity of being defended, i.e. a formal value.

9.3.2.4 The Dialogical System DLLC₂

We define the Dialogical Logic for Legal Condition **DLLC₂** once the following rules stated:

- *Particle Rules* = PR-SC ∪ PR-MO ∪ PR-DO ∪ PR-J ;
- *Structural Rules* = SR-0 ∪ SR-1 ∪ SR-3 ∪ SR-**O** ∪ SR-A

These sets of rules must be used in a restrict dialogical frame, i.e. determined by the sets $\mathscr{F}$ and $\mathscr{N}$. These two sets determine the conditions for the material dialog $\mathbb{DM}$ such that $\mathbb{DM} = \mathscr{F} \cup \mathscr{N}$.

Definition 3 (DLLC₂) **DLLC₂** is defined by the union of sets *Particle Rules* and *Structural Rules* used in $\mathbb{DM}$:

$$\mathbf{DLLC_2} = \frac{PartRules \cup StrucRules}{\mathbb{DM}}$$

9.3.3 Jérôme C. Guilt and Further Discussions

Since the formal details are now presented, we can use them to explore our Jérôme C. example. Subsequently, we will discuss some properties of our dialogical frame underlined by the example.

9.3.3.1 Jérôme C.'s Example

In our Jérôme C. example we have used the norm "Tax evasion must be punished". We reconstruct this norm via the logical rule $T_x \rightarrow \mathbf{O}P_x$. Whereas the set $\mathscr{N}$ doesn't change, we do change the set $\mathscr{F}$. Let us for now assume that "Jérôme C. has commited tax evasion" is a fact admitted by the Judge, as T_c (Table 9.8), and then let us consider that T_c is not admitted by the Judge and therefore does not belong to the set $\mathscr{F}$ (Tables 9.10 and 9.11). As we will see, when T_c does not belong to the set $\mathscr{F}$, the distinction between the two modes of discourse arises (Table 9.7).

In Table 9.8, Proponent adds T_c in the list $\mathscr{A}$ in move 2. In the next move, Opponent uses the structural rule **SR-A** to compel Proponent to utter this proposition in the contextual point 1, what he does in move 4. Then, Opponent requires a justification for the proposition T_c. Proponent manages to justify himself without difficulty, since T_c belongs to the set of facts (move 6). Once T_c is a fact belonging

Table 9.7 Initial concessions
for the dialog Table 9.8

$\mathscr{N}$	$\mathscr{F}$
$T_x \rightarrow \mathbf{O}P_x$	T_c

Table 9.8 Jérôme C. commits a tax evasion

O				P		
	$m := 1$				$\epsilon\|1 : [T_c]\mathbf{O}P_c$	0
					$n := 2$	
1	$\epsilon\|1: ?_{[]}$	0			$T_c\|1 : \mathbf{O}C_a$	2
3	$\epsilon\|1 : !_{(T_a)}$	2			$T_c\|1 : P_c$	4
5	$\epsilon\|1 : !_{[\![T_a]\!]}$	4			$T_c\|1 : [\![T_c]\!] \in \mathscr{F}$	6
7	$T_c\|1 : ?_2$	2			$T_c\|1.2 : P_c$	14
9	$T_c\|1 : T_c \to \mathbf{O}P_c$			$\mathscr{N}$	$T_c\|1 : ?_{c/x}$	8
11	$T_c\|1 : \mathbf{O}P_c$			9	$T_c\|1 : P_c$	10
13	$T_c\|1.2 : P_c$			11	$T_c\|1 : ?_2$	12

Table 9.9 Initial concessions
for dialogs Tables 9.10
and 9.11

$\mathscr{N}$	$\mathscr{F}$
$T_x \to \mathbf{O}P_x$	$\emptyset$

to the list of announcements, Proponent can request the legal order by asking for the particular norm created by the Judge with respect to fact T_c (move 8). Accordingly to this particular norm, Proponent manages to defend that it is true that Jérôme C. has committed tax evasion, and then it is obligatory that he has to be punished (move 14).

Now we consider what would happen if T_c is not an admitted fact (Table 9.9).

In Table 9.10, Proponent chooses to not commit himself with the defence of T_c and he defends with $\neg A_a$. Opponent can then only challenge the negation (moves 2–3). After this Opponent's challenge, Proponent has the choice:

1. He can require for a justification of T_c, or
2. He can use Opponent's challenge to change his previous defence.

In Table 9.10 we suppose that Proponent makes the choice 1. We develop the possibility of the choice 2. in Table 9.11.

Choice 1

After Opponent's move, Proponent requires a justification for T_c (move 4). Since T_c does not belong to the set $\mathscr{F}$, Opponent cannot use the propositional justification. Unfortunately for Opponent, T_c does not belong to any Proponent's previous moves. So, he can't use the procedural justification neither. Consequently, Opponent loses the play. Without proving Jérôme C.'s culpability, no particular norm could or could have been created by the Judge.

Table 9.10 Choice 1 – indicative tense

	O			**P**	
	$m := 1$			$\epsilon\|1: [T_c]\mathbf{O}P_c$	0
1	$\epsilon\|1: ?_{[\,]}$	0		$n := 2$	
3	$\epsilon\|1: T_c$	2		$\epsilon\|1: \neg T_c$	2
	–		3	$\epsilon\|1 : !_{[\![T_c]\!]}$	4

Table 9.11 Choice 2 – past or present conditional tense

	O			**P**	
	$m := 1$			$\epsilon\|1 : [T_c]\mathbf{O}P_c$	0
1	$\epsilon\|1: ?_{[\,]}$	0		$n := 2$	
3	$\epsilon\|1 : T_c$	2		$\epsilon\|1 : \neg T_c$	2
				$\otimes$	
5	$\epsilon\|1 : !_{(T_c)}$	4		$T_c\|1 : \mathbf{O}P_c$	4
7	$T_c\|1 : !_{[\![T_c]\!]}$	6		$T_c\|1 : T_c$	6
9	$T^*_c\|1 : ?_2$	4		$T_c\|1 : [\![T_c]\!] \in \mathscr{C}_{\mathbf{O}}$	8
11	$T^*_c\|1 : T_c \rightarrow \mathbf{O}P_c$			$T^*_c\|1.2 : P_c$	16
13	$T^*_c\|1 : \mathbf{O}P_c$		$\mathscr{N}$	$T^*_c\|1 : ?_{c/x}$	10
15	$T^*_c\|1.2 : P_c$		11	$T^*_c\|1 : T_c$	12
			13	$T^*_c\|1 : ?_2$	14

Choice 2

Opponent's challenge in move 3 allows Proponent to change his defence and then add T_c in the list $\mathscr{A}$ (move 4). Opponent uses the rule **SR-A** to compel Proponent to utter T_c in the contextual point 1 (moves 5–6). In the next move he requires a justification for T_c. Proponent cannot use the propositional justification, since T_c does not belong to $\mathscr{F}$. Nevertheless (contrarily to Table 9.10), the proposition T_c belongs to a previous move of Opponent. He had uttered T_c when he had challenged the negation (move 3). Regarding this move, Proponent has not required a justification, so he confers himself the possibility to reuse this proposition under the same conditions, and to justify it through a procedural justification. After this procedural justification, T_c gets marked by a * in the list, manifesting the discourse's changing of mode. The progress of the play is then based on what would or could have been the case once T_c was an accepted fact. In move 10, Proponent asks for the particular norm that the Jude would or could have created if T_c was a fact. The rest of the play is similar to what is developed in Table 9.8, excepting the value of the players discourse: they no longer debate over what is true, but rather over what could be or could have been true. The discourse is developed independently of the facts. It deals with the general norm's validity once its possible particularization, doing as if the fact was a verified and accepted one.

9.3.3.2 Further Discussions

The comparison between choices 1 and 2 have an important didactic aspect. Let us consider the object of the discussion between Proponent and Opponent. Proponent presents the following argument: "if it is proven that Jérôme C. committed tax evasion, then the equivalent sanction ought to be applied". But, before entering in this discussion, Opponent and Proponent agree to recognize that it is not yet admitted that Jérôme C. committed tax evasion. Consequently, Proponent only reaffirms that if Jérôme C. has committed tax evasion (which he is or could have been the suspect), it would be obligatory that he ought to be punished. The fact of his culpability does not change the fact of the obligation of the sanction in the case of culpability. Based on $\mathscr{F}$, Proponent shows that it cannot be obligatory that Jérôme C. is punished if it is not established that he is guilty of tax evasion (Table 9.10). Using the procedural justification, it's reaffirmed the dependence between "being obligatory" and the satisfaction of the condition (i.e. the validity of the general norm). This aspect manifests the imputation aspect linking the condition (have committed tax evasion) to the "ought" element (the sanction: to punish the behaviour). If the condition is not fulfilled, the "ought" concerning the behaviour cannot take place. The tax evasion is the condition for the "ought", for the sanction, but the sanction is conditional because it rises from a modality. If the latter is not itself fulfilled, Proponent wins, but only because the conditional relation is not invalidated by the non-satisfaction of the condition. Thus, the norm remains valid independently of the fact that Jérôme C. has not committed tax evasion.

Back to the Evidence of the Illegal Act

If the "ought" is associated with a condition, this condition has a particular character. It consists in the satisfaction of the particular norm's antecedent, what ensures the formulas added to the list $\mathscr{A}$. Remember that these formulas have a particular status. When a player changes the situation of the argumentative process (by choosing a new contextual point), due to the rule **SR-A** his adversary can compel him to (re)utter and then (re)justify them in this chosen situation. Consequently, since a formula is introduced to the list $\mathscr{A}$, a justification for this formula can be required in any new contextual point. When players are incapable of producing this justification, the particular norm cannot be obtained in this contextual point. Hence the punishment could not be said to be obligatory in it.

In order to achieve the sanction, the guilt has not only to be a fact, but especially it must be proved, that's to say justified in any situation. It could not exist a situation in which the sanction is pronounced and where the guilt cannot be justified. Jérôme C. cannot be punished for a tax evasion that he has not really committed. But this doesn't mean that he should never be punished at all: only that, while in absence of a proof of his guilt, no sanction can be applied. Even if everybody knows that he is guilty, that's not sufficient for the sanction to have to be applied. It must be proven that Jérôme C. is guilty, and the formalism surrounding the dynamic operator explicitly allows this. Moreover, it remains possible for Jérôme C. to commit tax evasion without becoming a suspect of such an act. For this, it's enough that we cannot justify this fact in any situation.

For a sanction to be applied to a guilty behaviour, we need that an authorized person (the Judge) creates a particular norm from a more general one, but also that this behaviour is proven.[48]

Remark

Deontic logic, in its treatment of legal norms, considers them as instantiations of general norms. The problem with this approach is that it reduces the normative propositions to conditional instantiations,[49] what addresses deontic logic to paradoxes linked to the material conditional. In our reconstruction of the legal reasoning, our proposal also deals with the (material) conditional to formalize the general norms. But the use of those norms presupposes the creation of a particular norm (to be created by the Judge only). The use of this particular norm is itself dependent on the evidence of the antecedent, which can only be attained by the dynamic operator. Thus, the obligation of the consequent is required only when the antecedent is true or procedural justifiable. Consequently, due to this formalization, not only the falsity of the antecedent cannot lead to the trivialization of the conditional relation, but also it allow us to study the meaning of a norm through its description, in order to give the conditions of its use when the fact is missing (procedural justification) or to give the truth conditions of the particular norm if the fact is the case (propositional justification).

9.4   Back to Jørgensen's Dilemma

Our original dialogical approach to logic is the most relevant frame to display the process concerning normative creation through an argumentative practice. We respect and preserve all the kelsenian theses with respect to the dualisms between the realm of "Is" and the realm of "Ought", between law and its science or theory, between a norm and its description, between the "legal actor"—the Judge, the Legislator, the authorized person—and the persons external to the legal procedures—the jurist, the legal scientist. Those elements play a specific role in our study of the legal reasoning through the process of a dialog, and they are essential to show how mistaken is the notion of a practical syllogism, which lead to theoretical problems like Jørgensen's dilemma.

---

[48]Even if the set $\mathscr{F}$ contains $\neg T_c$ or doesn't even contain T_c, identical plays will be produced. Maybe this indicates a link between the absence of the guilty proof and the innocence presumption.

[49]This is precisely the context where Jørgensen's Dilemma emerges.

Regarding the Dilemma, Kelsen had pointed out the fact that to create a particular norm (in the conclusion of Jørgensen's practical syllogism) the creator has to have the power to do so. In the legal context, this power, this authorization, must come from the law. Legal norms are always created via procedures which are internal to law itself. No one from the outside is able to create a valid norm, even with the support of the existence of a more general norm with the correspondent content. This is a necessary, but not sufficient, step in the normative creation. The "authorization" aspect cannot be ignored, and can never be attained by someone who is outside from the legal sphere. Each one of us, as citizens, is able to put a norm in question, to evaluate it as fair or unfair, to consider if it should be or not applied to a determined case. But Kelsen emphasizes that those are all acts of thought that do not attain the legal level of the valid act of will, which could be produced by anyone but the Judge, or the authorized person.

With the dialogs, what happens is a procedure of justification after this act of will has taken place. Moreover, it is a procedure which allows considering what would be the case once other conditions were at stake. Proponent and Opponent have no legal authority, they only discuss about the legal procedure, once it has already been achieved. Certainly, this can clear up a lot of legal aspects such as mistakes or flaws to be revealed in the process, and that is all the interest of a legal theory.

Our approach fits perfectly with Kelsen's solution to Jørgensen's dilemma, saying that we are all capable of questioning, justifying and reasoning about norms. We can guess what would be the result (the conclusion on the misleading practical reasoning), i.e. what would "normally" be the content of the result of a legal decision, for example. But by no means are we able to state this "result" as being a valid norm. There is no way to trespass the inference barrier in the practical syllogism, because, there is no such a way for a syllogism to be a practical one: syllogisms are constructed by acts of thought, and practical decisions are made by acts of will. It is a mistake to accept that we can simply mix up the two, or translate one into another, as it is a mistake to consider that norms can result of acts of thinking.

Our dialogic approach shows a perfect example of an application of Kelsen's theory concerning the important question of the normative creation. It displays how is it possible to confront oneself to general norms like "Tax evasion must be punished" and particular norms like "Jérôme C. (who committed tax evasion) must be punished" from a completely external point of view, and the justification element is central in this undertaking. We preserve Kelsen's temperance about the modally indifferent substract, yet showing how far can we go in analysing the different possibilities of interpretation. We show that it is perfectly acceptable to analyse, theorize, argue, examine, debate about norms, without transgressing the limits between the normative (practical) and the descriptive (theoretical) levels. No harming puzzles arise from this.

Appendix: Propositional Dialogic in a Nutshell

This Appendix presents the rules of Propositional Dialogic, i.e. the particle rules for the propositional connectors $\neg$, $\wedge$, $\vee$, $\rightarrow$ and the standard structural rules.
How should the particle rules be read?

The reading of the particle rules is straightforward once we keep in mind the notion of usage of the logical constant that they represent. A particle rule can be decrypted via these three points[50]:

1. An **X**'s utterance,
2. A challenge, which is the demanding made by player **Y** over the initial **X**'s utterance,
3. A defence, corresponding to the answer of player **X** to the challenge made by **Y**.

Particle Rules

Before presenting the structural rules, we have to deliver one more definition: that of repetition rank. A repetition rank is a positive integer corresponding to the number of times that a player can repeat the same challenge or the same defence.[51]

Table 9.12 Standard connectives – **PR-SC**

Burden and/or object of choice	X-Utterance	Y-Challenge	X-defence de X
$\neg$, there is no defence	$\mathscr{A}\|i : \neg\varphi$	$\mathscr{A}\|i : \varphi$	$\otimes$
$\wedge$, the challenger has the choice	$\mathscr{A}\|i : \varphi \wedge \psi$	$\mathscr{A}\|i : ?_{\wedge 1}$ *or* $\mathscr{A}\|i : ?_{\wedge 2}$	$\mathscr{A}\|i : \varphi$ *respectively* $\mathscr{A}\|i : \psi$
$\vee$, the defender has the choice	$\mathscr{A}\|i : \varphi \vee \psi$	$\mathscr{A}\|i : ?_{\vee}$	$\mathscr{A}\|i : \varphi$ *or* $\mathscr{A}\|i : \psi$
$\rightarrow$, both players shared the burden	$\mathscr{A}\|i : \psi \rightarrow \varphi$	$\mathscr{A}\|i : \psi$	$\mathscr{A}\|i : \varphi$

---

[50]Excepted the rule for negation since there is no defence, see Table 9.12.

[51]See [3, chap. 2] for a further discussion on repetition ranks.

Structural Rules

◊ **Starting Rule SR-0**: Any play d_Δ of a dialog $\mathscr{D}_\Delta$ starts with **P** uttering Δ— the thesis. After the utterance of the thesis, **O** has to choose a repetition rank. **P** chooses his repetition rank right after **O**.

◊ **Playing Rule SR-1**: Players move alternatively. Each following the repetition rank is either a challenge or a defence concerning a previous challenge.

◊ **Atomic Restriction SR-2**: **P** cannot utter an atomic formula first. He is only allowed to reused those previously uttered by **O**.

◊ **Winning Rule SR-3**: A player **X** wins a play if and only if it is **Y**'s turn play but he cannot move anymore with respect to the rules.

Acknowledgements This paper has been supported by JuriLog Project (ANR11 FRAL 003 01), hosted at the Maison européenne des sciences de l'homme et de la société (MESHS – USR 3185).

References

1. C.A. Cabrera, Imperativos y lógica en Jørgen Jørgensen. Isegoría **20**, 207–215 (1999)
2. V. Champeil-Desplats, Alf ross: droit et logique. Droit et société **1**, 29–42 (2002)
3. N. Clerbout, Étude de quelques sémantiques dialogiques. Concepts fondamentaux et éléments de métathéorie. PhD thesis, Université de Lille, 2013
4. H. van Ditmarsch, W. van der Hoek, B. Kooi, *Dynamic Epistemic Logic*. Synthese Library: Studies in Epistemology, Logic, Methodology, and Philosophy of Science, vol. 337 (Springer, Dordrecht, 2008)
5. J. Hansen, G. Pigozzi, L. van der Torre, Ten philosophical problems in deontic logic. Norm. Multi-agent Syst. **7122**, 55–88 (2007)
6. R.M. Hare, *The Language of Morals* (Oxford University Press, Oxford, 1972)
7. L. Keiff, Dialogical logic (2009), http://plato.stanford.edu/entries/logic-dialogical
8. H. Kelsen, *General Theory of Norms* (Oxford University Press, Oxford, 1979). Re-edited in 2011
9. P. Lorenzen, K. Lorenz, *Dialogische logik* (Wissenschaftliche Buchgesellschaft, Darmstadt, 1978)
10. S. Magnier, PAC vs. DEMAL, a dialogical reconstruction of public announcement logic with common knowledge, in *Logic of Knowledge. Theory and Applications*, ed. by C. Barès, S. Magnier, F. Salguero (College Publications, London, 2012), pp. 159–179
11. S. Magnier, *Approche dialogique de la dynamique épistémique et de la condition juridique* (College Publications, London, 2013)
12. S. Magnier, T. de Lima, A soundness & completeness proof on dialogs and dynamic epistemic logic, in *Dynamics in Logic*, ed. by P. Allo, F. Poggiolesi, S. Smets. Forthcoming as special issue of Logique et Analyse (2015)
13. S. Rahman, Non-normal dialogics for a wonderful world and more, in *The Age of Alternative Logics*, ed. by J. van Benthem, G. Heinzmann, M. Rebuschi, and H. Visser (Springer, Dordrecht, 2002), pp. 311–334
14. S. Rahman, H. Rückert, Dialogische Modallogik (für T, B, S4, und S5). Logique et Analyse **167**(168), 243–282 (1999)

15. S. Rahman, T. Tulenheimo, From games to dialogues and back, in *Games: Unifying Logic, Language, and Philosophy*, ed. by O. Majer, A. Pietarinen, and T. Tulenheimo (Dordrecht, Springer, 2006), pp. 153–208
16. A. Ross, Imperatives and logic. Philos. Sci. **11**(1), 30–46 (1944)
17. G. Sartor, *Legal Reasoning: A Cognitive Approach to the Law*. Treatise on Legal Philosophy and General Jurisprudence, vol. 5 (Springer, Berlin, 2005)
18. G.H. von Wright, Deontic logic. Mind **60**(237), 1–15 (1951)
19. G.H. von Wright, *Logical Studies* (Routledge and K. Paul, London, 1957)

Chapter 10
Note on a Second Order Game in Legal Practice

Rainhard Z. Bengez

Abstract In this contribution we investigate the structure of judgment aggregation at courts of lay assessors. After we have demonstrated how to avoid paradoxes related to the propositional logical structure of judgment aggregation we introduce a game theoretical framework, the P-game. This game proves successfully to illustrate the process of practical decision making among lay judges and regular judges; furthermore it opens a fruitful theoretical framework for deeper investigations into negotiating strategies related to courts of lay assessors.

10.1 Introduction: Back Room Dispute and the Wheel of Power

In a paper on juridical decisions, Philipps [1] demonstrated colorfully how the problem of *multiple non-transitive preference relations*[1] regularly occurs in German courts of lay assessors. As a possible solution he outlined to transform such conflicting preference relations into a scheme consisting of cardinal numbers[2] to find a common agreement. His strategy is put into effect in two steps:

1. Converting the preference relations into a metrical form
2. Dialog and negotiation[3]

---

[1]The preference relation we have in mind is whether a given crime or criminal is more *punishable* than another one. For example, if we compare three opinions we will have three different evaluations, i.e. three differently ordered degrees of penalty. The problem the court has to solve is now to transform all these different opinions into a common judgment.

[2]In a legal context such cardinal numbers are represented by the intensity of punishment or duration of fines.

[3]De facto he showed how to apply a specific kind of legal dialogical logic to find an agreement (or to convince the opponents from someone's position in the one or other way). It is important to point out that this negotiation is *now* related to cardinal numbers and to the question how the opponents

R.Z. Bengez (✉)
MCTS, TU München, Carl von Linde Akademie Arcisstr. 21, 80333 Munich, Germany
e-mail: bengez@cvl-a.tum.de

© Springer International Publishing Switzerland 2015   223
M. Armgardt et al. (eds.), *Past and Present Interactions in Legal Reasoning and Logic*, Logic, Argumentation & Reasoning 7, DOI 10.1007/978-3-319-16021-4_10

This double strategy avoids running into one of the paradoxes associated with *judgment aggregation*. Judgment aggregation is a relatively young field of research investigating the propositional nature of multiple judgments aiming at finding a common agreement. This discipline is interpreting the structure of joint decisions in terms of propositional logic, i.e. means firstly to formulate the problem in terms of a logical proposition; and secondly to associate to each logical sentence a truth value.[4] In the light of two valued logic each singular judgment is associated with a single truth value. The problem occurring now is that given a sentence we can have two distinct evaluations, i.e. one judgment labels the sentence as *true* and the second judgment marks the sentence as *false*. In terms of any logic relying on the *tertium non datur* concept it is hard to find a common judgment. If we use any kind of propositional logic, then it is easy to understand that we could formulate so called impossible theorems expressing that under certain formal circumstances it is not possible to find a joint agreement. A real world problem is that any real court has not only to find a solution, but to agree in time. To complete this challenging time and fairness related task legal traditions have developed several strategies.

At this stage Lothar Philipps's double strategy comes into play. This strategy presupposes that the first round of negations between lay and regular judges ends when each judge has completed his preferred order which he is willing to defend. At this point they recognize that they will not easily find a common agreement and fear to run into one of the paradoxes of judgment agreement theory. Now, it is important to recall that any practical judgment does not aim at deciding whether one criminal is more punishable than another one, but to assign a degree of penalty[5] as punishment to each defendant. To start with the second round of the negotiation game we have to transform the *punishable*-relation and to increase the amount of information. To get more information we have to enrich our relation by going beyond its order structure. This can be done by substituting the ranks of the culprits with provisional degrees of penalty, i.e. the judges extend the order by introducing additional information, namely the years of lifetime they have to pay for their crimes. What did we gain? Firstly, we substitute the names of the defendants by simply writing down the years

---

are willing to bridge the gap between their solely numerical differences. Stated otherwise, it is easier to find an agreement whether someone shall pay a fine of 100\$ or 150\$ instead of agreeing whether A is more punishable than B. In the first case we are discussing about a difference of \$50 and will maybe agree that a fine of \$125 is fair because nobody or everyone loses. In the second case the setting is much harder because this situation is focused on identifying a unique winner and a unique looser. Thus, in the second case the positions will be defended very hard, and even if it is possible to find a solution then the process of negotiation is much more time consuming.

[4]In elementary first order propositional logic this would always be one of the dyadic values: *true or false*. Stated otherwise, a judgment aggregation based on elementary first order propositional logic or modal propositional logic relies on the bivalence principle (*tertium non datur*), i.e. that there are only two possibilities A and non-A, but no third one. Modal propositional logic is just an extension of the basic first order propositional logic.

[5]German penal code: § 39 para. 2 expresses that degrees of penalty have been allocated in years and month.

Table 10.1 Example of a Preferred order relation of punishment. The table should be read as follows: X > Y means that X is more punishable than Y

Judge	Preferred order
L_0	$D_0 > D_1$ and $D_2 > D_1$ and $D_2 > D_0$
L_1	$D_1 > D_0$ and $D_1 > D_2$ and $D_2 > D_0$
R	$D_0 > D_1$ and $D_1 > D_2$ and $D_0 > D_2$

and month of their individual punishment and we get a list of (ordered) numbers. Secondly, now we have numbers (years and months), so called cardinals; and from cardinals we easily can get differences.

Let us illustrate the entire situation. Let's say we have three defendants D_0, D_1 and D_2 as well as three judges – two lay judges L_0 and L_1 and one regular judge R. Each judge has a preferred order of how punishable each defendant is, i.e. an individual *punishable*-relation.

From Table 10.1 we can easily read off that we never get a clear overall transitive relation, i.e. something like $A > B > C$ which means that $A > B$ and $B > C$ and $A > C$. And, therefore, it is not possible to formulate a common or joint agreement in general. In our example, the only judge having such a relation is the regular judge R. What can we do to solve this imbroglio?

According to Philipps's strategy we have to transform the preferred order of punishment into degrees of penalty, i.e. we have to convert a mere ordinal order and its labels into cardinal numbers.

Formally spoken, we have to map from an arbitrary set into real numbers. To do this we proceed as follows:

(a) We start with an arbitrary, unordered set of defendants: S_{uo}
(b) According to our preferences and evaluation of facts and evidences F we choose two elements of S_{uo}: $x \in S_{uo} \times S_{uo} =: S_{uo}^2 =: S^2$
(c) The process of selection can be described as mapping from S_{uo} onto S^2 according to the preference F_1, whereas F_1 describes the preference related to the first selection: $F_1 : S_{uo} \rightarrow S^2$
(d) We have to repeat the previous step two times to compare each defendant with one other. More generally, we have to evaluate n defendants in $n \cdot (n - 1) / 2$ comparisons. In our example we have 3 defendants and need $3*2/2 = 3$ comparisons. If we had 5 defendants, we would need $5*4/2 = 10$ comparisons. This means that we need $n \cdot (n - 1) / 2$ mappings $F_{\_}$ from S_{uo} to S^2 for each judge. To make things a bit easier we combine all those F_k into one multi-dimensional mapping $F := (F_1, F_2, F_3, \ldots, F_{n(n-1)/2})$ to describe the entire selecting and comparison procedure.
(e) What we got now is: $F : S_{uo}^{n(n-1)/2} \rightarrow S^2$. And, as shown above, this is something which may lead us into troubles as we can't guarantee that we will get always a transitive relation of preferences. A transitive relation would mean that we can make a continuous chain out of every fragment any judge is

Table 10.2 Example of
degrees of penalty in years[a]

Judge	Degrees of punishment
L_0	$2_{D0} > 1_{D1}$ and $3_{D2} > 1_{D1}$ and $3_{D2} > 2_{D0}$
L_1	$3_{D1} > 1_{D0}$ and $3_{D1} > 2_{D2}$ and $2_{D2} > 1_{D0}$
R	$3_{D0} > 2_{D1}$ and $2_{D1} > 1_{D2}$ and $3_{D0} > 1_{D2}$

[a]The indices D_x are just illustrating the degrees of penalty's association with the preferred order. They are not necessary, but facilitate reading the table

contributing. This presupposes that we can connect each fragment of the chain with other fragments and remain nothing.

(f) In this step we need real numbers, i.e. a way to quantify names and its order. To put it in another way: let's say we have something like Mike > Tim & Jane > Jill & Tim > Jill & Jane > Mike and now we have to assign numbers to names preserving its underlying order. This sounds simple, but this quantification is quite hard to practice. In formal terms we are just assigning real numbers to our preferences $F: Q : S^2 \rightarrow \mathbb{R}^{n(n-1)/2}$.

In formal terms and by using a set-theoretic notion it is easy to formalize the entire process. To make it stronger we would have to characterize the underlying functions and relations, but this is not that necessary for our theory or scope.

In legal practice the entire procedure will be done by experience and negotiation, comparison and balancing. Maybe we get something like Table 10.2.

By assigning to each defendant a degree of penalty *additional information* has been put into the ordered list of defendants.[6] In other words, ordinal numbers have been transformed into cardinal numbers. Those numbers have one big advantage: they can be used for elementary calculations, e.g. we can subtract them and quantify the differences between each defendant's degrees of penalty. The open and interesting question is what kind of information has come into play, and how it has been used? What we have done, in philosophical terms, was that we have changed the concepts. Does this mean that there are *semantic(!)* concepts closer to numbers and quantification than others? And, how and when do we bridge this gap? What are the underlying proto-theories, epistemic or social ontologies or ontic structures justifying our approach?

It seems that we have done nothing spectacular, but what we have gained by transferring it into cardinal numbers is something great: we have omitted the paradoxes of judgment aggregation by leaving propositional logic behind us. By using Table 10.3 we can now start to dispute whether a specific degree of penalty

---

[6]Where does this additional information come from? Plastically spoken: from the judge's individual decision to assign concrete figures, i.e. years and month to a specific defendant. Compared with a simple order this is something new because he is not free to assign an arbitrary sequence of figures to his preference relation conserving its monotonic order. The selection of figures has to be justified and balanced between the defendants and their offence and within a socio-cultural framework.

Table 10.3 Degrees of punishment illustrating a stalemate situation

Judge	Defendant D_0	Defendant D_1	Defendant D_2
L_0	2	1	3
L_1	1	3	2
R	3	2	1

is justified. Right now it is easier to find a compromise and it is easier to cooperate. Again, a social or legal philosophical question, why is it easier to find a compromise after we have introduced and applied a kind of quantification? Furthermore, the structure of the underlying game has been changed in that way that the judges can agree sequentially. This means they can discuss specifically about each defendant's fair degree of penalty. It means we have cut a monolithic structure into slices onto which we can iterate an exchange. This is again a kind of quantification. Stated otherwise, by transferring it into cardinals we put additional information in the underlying structure. This additional information enables us now to reduce the complexity of the entire situation by looking at each defendant separately. In terms of complex systems, we have created a (linearly) combined system.

This is a great finding of Philipps [1] and what we have done so far was just to use his blueprint and his examples to introduce some new thoughts and to show some formal consequences. In the remaining contribution we will focus on the second part of the back room negotiation and introduce a new game which we will call the P-game.[7] The P-game starts with something similar to Table 10.3, i.e. with a restricted game based on cardinal numbers. In this situation, for example, the regular judge has to lead the negotiation to a final result.

10.2   Structure of the Second Game

In this section we will shortly introduce a new game, the P-game, within the framework of game theory. This game is based on the transformed propositional logical preference structure of juridical judgment aggregation. This new structure is composed of and based on cardinal numbers. As stated above, its complex structure can be seen as a collection of linear compartments. This enables us to play the P-game in two variations. The first variation is a double sequential game in which the regular judge R tries to negotiate the degree of penalty of each defendant individually with each lay judge, e.g. he starts to negotiate the degree of penalty of D_0 with L_0 and after they agreed he continues the negotiation with L_1 and so on. In a further variation of the P-game the negotiation about the degree of penalty of each defendant has to be done synchronically with all lay judges together, i.e.

---

[7]P-game abbreviates not only Philipps-game, but also the underlying idea that *Piscem vorat maior minorem [Latin for the big fishes swallow the smaller ones]* and our concept to better understand may change this situation slightly.

simultaneously. In this contribution we just introduce the first variation which makes extensive use of the linear complexity of the cardinal structure of the P-game.

10.2.1 Roles and Pay off Structure

Now, we start to identify the lay judges and the regular judge as opponents (and players). Each player can choose among two different strategies:

(a) Rejection
(b) Cooperation

In other words, the regular judge R starts to formulate a compromise between his degree of punishment concerning a specific defendant and specific position of his colleague (cf. for example Table 10.3). We can illustrate this situation as follows: After several minutes of discussion R writes down a number and hands over the note to L_x. L_x now has the possibility to accept or to reject his suggestion. After he has made his move this round is closed. For the beginning, let's say, we just have a single shoot game. Later we can extend this game to (in-) finitely many rounds. It is obvious that they can only win this game if they cooperate, i.e. just in that case they have found a sound trade-off between their quantified positions. In case, they will not find a compromise, i.e. in terms of our language both parties reject and, thus, both parties will lose. If just one party rejects the (written) suggestion, then the rejecting party wins. If someone loses, he gains nothing; therefore we can assign 0 and in case someone wins we can assign a positive value like 1 respectively as virtual trade-off. We can summarize this observation in Table 10.4.

Table 10.4 is helpful in a brief analysis of the game. The first thing we can read off is that the best strategy (from a bird's perspective) to finish the game is cooperation, but from an individual perspective the preferred strategy is to reject. If someone rejects, then the game continues and in the next turn he has to formulate a compromise. At this stage several questions arise: how can someone win this game and how long does it take or what is the best strategy to choose?

Table 10.4 Strategies and payoffs of the judges

	Player L (lay judge)		
Player R (regular judge R)	**Strategies**	*Rejection*	*Cooperation*
	Rejection	$(0_R, 0_L)$	$(1_R, 0_L)$
	Cooperation	$(0_R, 1_L)$	$(1_R, 1_L)$

10.2.2   A First Analysis of P-game

A second glance at the structure (cf. Table 10.4) of the P-game shows that this game has infinite number of equilibria. In other words, if an opponent tries to maximize his chance to make the game, then he has to take vagueness and uncertainty into account. A little bit more formally: If we assume that player R answers with strategy *rejection* with a yet unknown probability p and with strategy *cooperation* with probability 1−p under the condition player L chooses the strategy *rejection*, then we get the following expected trade-off:

$$\mu_{L,rejection}\left(R_{rejection,p.}\right) = 0_L \cdot p + 1_L \cdot (1-p) = 1-p \qquad (10.1)$$

In case the lay judge L_x always uses strategy *cooperation* the expected trade-off is:

$$\mu_{L,cooperation}\left(R_{rejection,p}\right) = 0_L \cdot p + 1_L \cdot (1-p) = 1-p \qquad (10.2)$$

Thus we can derive:

$$\mu_{L,rejection} = \mu_{L,cooperation} \qquad (10.3)$$

Because of equality (10.3) we can claim that there are indeed infinitely many values of probability *p* holding (10.3) and, therefore, a lot of mixed strategies are possible as the expected pay-off for both strategies *rejection* and *cooperation*, respectively, are equal. The meaning of this result is: if we are playing a single shot game the best strategy to come to a conclusion (i.e. agreement) would be cooperation. Stated otherwise, the lay judge would have to accept the regular judge's suggestion or vice versa. In case of repeated games, i.e. in case of a longer discussion and many more turns, and this will be the usual situation, the willingness to cooperate will decrease.

What we can learn from this first result is that we can formally confirm the feeling that a strong and experienced judge will dominate the back room discussion with his authority and closing it with his preferred result.

10.2.3   A Deeper Analysis of the P-game

In the last section we have learned that from a meta-perspective cooperation would be the best choice in terms of costs (time, money, welfare, happiness, etc.), i.e. the game is coordinated. In case, we are playing a repeated P-game the degree and willingness of cooperation decreases. This can be illustrated by the following payoff matrices:

$$P = \begin{pmatrix} 0 & 1 \\ 0 & 1 \end{pmatrix} \qquad (10.4)$$

Table 10.5 Generalization of Table 10.4: introducing variables instead of concrete payoff values

	Player L (lay judge)		
Player R (regular judge R)	**Strategies**	*Rejection*	*Cooperation*
	Rejection	(a_R, a_L)	(b_R, c_L)
	Cooperation	(c_R, b_L)	(d_R, d_L)

Table 10.6 Generalization of Table 10.5: omitting indices. This points the symmetry properties of the payoff structure of P-game out[a]

	Player L (lay judge)		
Player R (regular judge R)	**Strategies**	*Rejection*	*Cooperation*
	Rejection	(a,a)	(b,c)
	Cooperation	(c,b)	(d,d)

[a]The content of Tables 10.5 and 10.6 can be led back to Table 10.4 by the following transformation: $a = 0, b = 1, c = 0, d = 1$. What we did in the transition from Tables 10.5–10.6 was to say $a_R = a_L$ and this we abbreviate by a, i.e. $a = a_R = a_L = a$

$$P^T = \begin{pmatrix} 0 & 0 \\ 1 & 1 \end{pmatrix} \qquad (10.5)$$

A further observation is that we can't discover any complementary relation as it is many times observed in coordination games. A coordination game describes a family of negotiations in which player may choose the same or a corresponding strategy. This family or class of games is widely used to describe social actions where the agents are depending on each other like finding the best road to the university or cinema. Here, the costs i.e. the duration to drive to the cinema or university depends on the number of drivers using a specific road. Coordination games are not cooperation games, i.e. the players or agents don't have to talk and to work together to bring their strategy to action. A game is complementary, if the added value gained by player X increases the added value of player Y, too; a so called win-win-situation. It seems obvious that in P-games there should be no added value between the players expect a sound and fair judgment related to the case and a less time consuming negotiation. This is because it describes a negotiation between judges and is related to justice and the legal system. Therefore, it is strongly connected to moral systems and ethical questions. To illustrate this we need the formalizations of Tables 10.5 and 10.6.

What we do now is a proof by contradiction. This means we *have to assume that our P-game is(!)* a complementary game. In other words: $c > a, d > b$ and $b > c$ & $d > a$ holds as well. This leads to the following equations:

In case lay judge's L_x strategy is 1, then judge R can switch from strategy 1 to strategy 2. We can write the gained added value related to this switch as:

$$added\ value\ of\ R : {}^{L:1}_{R:1 \to 2} = c - a > 0 \qquad (10.6)$$

In case lay judge L_x adopts strategy 2, then the same switch as in Eq. 10.6 will provide judge R a higher added value as:

$$added\ value\ of\ R\!:^{L:2}_{R:1\to2} > added\ value\ of\ R\!:^{L:1}_{R:1\to2} \qquad (10.7)$$

This holds because of:

$$added\ value\ of\ R\!:^{L:2}_{R:1\to2} = d - b > c - a = added\ value\ of\ R\!:^{L:1}_{R:1\to2}.$$

This is something one can derive by applying the assumptions. In other words, we make use of the added value concept or welfare which is ordered from (d,d) to (a,a), i.e. from the highest to the lowest values.

Now we got a contradiction as in our P-game the payoff structure is different, i.e. d–b = 0 and c–a = 0. Thus, Eq. (10.7) does not hold. In the contrary, for us holds:

$$added\ value\ of\ R\!:^{L:2}_{R:1\to2} = added\ value\ of\ R\!:^{L:1}_{R:1\to2} \qquad (10.8)$$

Last but not least, a coordination game has sometimes a further property called *positive spillovers*. This means, that each player can increase his own added value when the other player switches his strategy. Do positive spillovers exist in our P-game? Yes, as if, e.g. lay judge L_x switches his strategy judge R can increase his added value:

$$added\ value\ of\ R\!:^{L:2\to1}_{R:1} = b - a > 0 \qquad (10.9)$$

$$added\ value\ of\ R\!:^{L:2\to1}_{R:2} = d - c > 0 \qquad (10.10)$$

$$added\ value\ of\ R\!:^{L:1\to2}_{R:1} = 1 \qquad (10.11)$$

$$added\ value\ of\ R\!:^{L:1\to2}_{R:2} = 1 \qquad (10.12)$$

We can conclude our second analysis with the knowledge that P-game is a coordination game with positive spillover but without complementary relation.

10.3   Final Remarks and Outlook

We left many things open, e.g. strategies how to win a repeated game, its evolutionary character and an empirical study about its relevance, but on the other hand we got a deeper insight into the structure of back room negotiations.

The first thing we could demonstrate was that the transformation of ordered structures (within a legal context) into cardinal numbers avoids running into the trap of propositional paradoxes. Secondly, we have identified the game theoretical character of the decision finding process in courts of lay assessors and introduced the structure of a new game, the so called P-game and its associated strategies. In the basic case of a non-repeated game we further could show that the best strategy would be accepting the first compromise. In case of repeated games this result won't hold as the expected pay-off of cooperating and rejecting are identical (cf. equality (10.3)). Mixed strategies are not possible in the perturbed P-game which is a coordination game; but are they possible in ordinary P-game? What could be a proper strategy to win this game as lay judge? An idea could be to force the partner to switch as we have shown that switching of the opponent increases someone's own added value. In forthcoming works we will further investigate into the structure of P-game and demonstrate an empirical study.

What we can learn from this first result is that we can formally confirm the feeling that a strong and experienced judge will dominate the back room discussion with his authority and closing it with his preferred result. What we already know is that the regular judge dominates the scene. This is something we can explain in terms of the P-game. Furthermore, we know that the discussion will find an end as there is nothing to gain by a longer dispute, i.e. the game does not show a complementary relation. If someone is able to convince or persuade his discussion partner to change the strategy, e.g. by providing a compromise based on the quantification transformation, then he will make the game. This is what we learned from the previous section as P-game shows positive spillovers.

Addendum 1: Impossible Theorem of Joint Agreement

In the first section we claimed that in propositional logic it is not possible to find a joint agreement. In other words, by omitting quantification it is not possible to find a compromise if certain conditions hold. What we claim is, given that two judges differ in one evaluation, i.e. A says X is guilty and B says that X is not guilty or that X is more punishable than Y and B says that Y is more punishable than X. In this case it is not possible that they will find a common agreement. This is quite clear as a common agreement would mean that they have in every singular atomic proposition or its logical connection the identical evaluation. The core of the proof is the evaluation function, i.e. their preferences. In a logical setting we would have to check whether they assign to elementary propositions identical truth values. If not, and in case we are using any kind of two valued and *tertium non datur* related logic, we would get something like W and *non-W*. And these two differently evaluated propositions can never be part of a joint agreement, i.e. a common and shared evaluation function.

The Table 10.7 illustrates this situation.

As there is neither a single row with identical evaluations a joint agreement and unique judgment by a mere application of propositional (or first order) logic is not possible. But, as we know, here comes quantification into play.

Table 10.7 Shared evaluation functions

	Evaluation/preference of Judge 1	Evaluation/preference of Judge 2	Evaluation/preference of Judge 3
Evidence 1	Yes	Yes	No
Evidence 2	Yes	No	Yes
Evidence 3	No	Yes	Yes

Addendum 2: Transformation of P-game into a Coordination Game

In the previous chapter we mentioned several times that P-game is a coordination game and even showed that is has the property of the positive spillover. Until now we owe a proof. How can we transform P-game into a coordination game? This can be done introducing a perturbation parameter P.

$$added\ value\ of\ R{:}_{R:1\rightarrow2}^{L:2} = d - b = 1 - 1 = 0 \tag{10.13}$$

$$added\ value\ of\ R{:}_{R:1\rightarrow2}^{L:1} = c - a = 0 - P = -P \tag{10.14}$$

Therefore, we get:

$$added\ value\ of\ R{:}_{R:1\rightarrow2}^{L:2} > added\ value\ of\ R{:}_{R:1\rightarrow2}^{L:1} \tag{10.15}$$

This reformulation of P-game requires slightly different payoff matrices (see Tables 10.8 and 10.9).

What we can easily see is that it is not possible to construct a mixed strategy in a reasonable way as it would contradict Eq. 10.3 or require that $P = 1$ as one can derive from the following equations (Table 10.10):

$$\mu_{L,rejection}\left(R_{rejection,p,}\right) = P \cdot p + 1 \cdot (1 - p) = 1 - (1 - P) \cdot p \tag{10.16}$$

Table 10.8 P-game with Perturbation Factor P. The selected strategy is *Reflection*

	Player L (lay judge)		
Player R (regular judge R)	**Strategies**	*Rejection*	*Cooperation*
	Rejection	**(0 + P,0 + P)**	(1,0)
	Cooperation	(0,1)	**(1,1)**

Table 10.9 P-game with Perturbation Factor P. The selected strategy is *Cooperation*

	Player L (lay judge)		
Player R (regular judge R)	**Strategies**	*Rejection*	*Cooperation*
	Rejection	**(0,0)**	(1−P,0)
	Cooperation	(0,1−P)	**(1,1)**

Table 10.10 Generalization of Tables 10.8 and 10.9 to follow the proof easily that perturbed P-game is a coordination game

Player R (regular judge R)	Player L (lay judge)		
	Strategies	*Rejection*	*Cooperation*
	Rejection	**(A,a)**	(C,c)
	Cooperation	(B,b)	**(D,d)**

$$\mu_{L,\,cooperation}\left(R_{rejection,\,p}\right) = 0 \cdot p + 1 \cdot (1-p) = 1 - p \qquad (10.17)$$

Now we can say that the perturbed P-game is a coordination game as $A > B$ (Table 10.7), $D > C$ (Table 10.8) holds for the judge R and $a > c$ (Table 10.7), $d > b$ (Table 10.8) holds for the lay judge L. The bold marked fields in the tables represent so called Nash-equilibria.

Reference

1. L. Philipps, Das Abstimmungsparadoxon im juristischen Alltag. Slovenian Law Rev. **2**, 33–39 (2005). Re-printed in: L. Philipps, *Endliche Rechtsbegriffe in unendlichen Grenzen: Rechtslogische Aufsätze* (Editions Weblaw, Bern, 2012), pp. 119–125

Chapter 11
Conflict of Norms and Conflict of Values in Law

Sandrine Chassagnard-Pinet

Abstract The conflict of norms is characterized by the existence of a rule that prescribes something while a second, also valid rule, prescribes the opposite. Thus, the presence of antinomic norms undermines the belief in a logical unity of the legal order. For this reason, the existence of antinomies in law is widely discussed, as well as the cause and the nature of these conflicts. Often imputed to a failure in the norm producing process, these conflicts of norms are treated as a pathological phenomenon that we should try to avoid at the source or to interpret away by logical means. However, these preventive or curative tools do not address and solve the antinomies which are located at the deeper level of the values hiding behind the legal rules and supporting them. This article discusses the need for a classification of different kinds of normative conflicts and, thus, the need for appropriate strategies to solve them.

11.1 Introduction

According to Portalis [24], one of the contributors to the French Civil Code, "it would be a great evil if there were to be contradictions between the maxims that govern men".[1] Since Portalis understood codified French law as a complete and consistent normative system, without shortcomings or omissions, it was difficult for him to admit the presence of antinomies in law and he could only conceive them as an abnormal phenomenon, the signs of a pathology of law. Because they go against the lawmaker's rationality postulate (see Ost [20, pp. 97 ssq]), normative conflicts are often considered to be the result of a failure in the norm producing process. According to this view, the appearance of antinomies would be a consequence

---

[1] Original text: "Ce serait un grand mal qu'il y eût de la contradiction dans les maximes qui gouvernent les hommes".

S. Chassagnard-Pinet (✉)
Full Professor of Private Law, Equipe René Demogue (CRD&P, EA 4487), Université de Lille (UDSL), Lille, France
e-mail: sandrine.chassagnard-pinet@univ-lille2.fr

© Springer International Publishing Switzerland 2015   235
M. Armgardt et al. (eds.), *Past and Present Interactions in Legal Reasoning and Logic*, Logic, Argumentation & Reasoning 7, DOI 10.1007/978-3-319-16021-4_11

of the abundance of norms as well as of the lack of coordination between them. Contradictions appear if the author of a new norm has failed to ascertain whether it is compatible with already existing ones. However, beyond this interpretation of conflicts of norms, which reduces the contradictions between legal rules to mere legal defects, it is possible to look for more substantial causes hiding behind these antinomies and consider them as symptoms of existing tensions at the deeper level of legal foundations. From this perspective, normative antinomies point to the existence of conflicts between the values which inspired the norms [14]. Indeed, these are the cases in which the interpreter has the greatest difficulties in resolving the conflicts of norms. As Perelman [23, p. 36] noted, "one of the main tasks of legal interpretation is to find solutions to conflicts between rules by means of establishing a hierarchy of values, which these rules are meant to protect".[2]

Legal science has at its disposal different logical principles to connect different norms – which led some legal scholars to adopt the view that every conflict may be interpreted away and that the legal system does not harbour any actual antinomy. However, these interpretation techniques are not in the position to deal with the conflicts between values that underlie some normative antinomies. In such cases, the judge cannot rely anymore on techniques of logical interpretation. Instead, he has to turn to a pragmatic approach resting on his assessment of existing interests. Far from limiting himself to play the role of "the mouth that pronounces the words of the law" [18, p. 163], as the exegetical school would like [25], the judge becomes normatively active and proposes an interpretation that makes norms compatible in consideration of the values they convey. Indeed, as Amselek [1, p. 499] puts it, it is up to "the interpreters to find remedies, to overcome contradictions, to search for solutions through an appropriate excavation of the sense of a norm – all of which should put the applying organs in the position to deliver their decisions or the legal subjects in the position to determine whether their behaviour is law-obeying".[3] Interpreters, insofar as they act as "necessary reducers of uncertainties"[4] (*Ibid.*) have to struggle with the logical deficiencies of norms to render their application possible.

Because it goes against the unity and consistency of the law, and because it jeopardizes the reliability and foreseeability essential to legal regulations, the existence of legal antinomies, in so far as it is acknowledged, forces the judge to develop methods to deal with them. Thus, we will have to appreciate the role played by logical arguments within these strategies. The preliminary task will be to assess

---

[2]Original text: "Une des principales tâches de l'interprétation juridique est de trouver des solutions aux conflits entre les règles, en hiérarchisant les valeurs que ces règles doivent protéger".

[3]Original text: "Aux interprètes d'apporter des remèdes, de surmonter les contradictions, en allant rechercher par des approfondissements de sens appropriés des solutions qui permettront aux organes d'application de rendre leurs décisions ou aux sujets de droit de fixer leurs comportements d'observance".

[4]Original text: "Des réducteurs obligés d'incertitude".

whether legal antinomies do indeed occur in law. Then, we turn to the possibility of distinguishing between different types of antinomies, which could lead to specific ways to deal with them.

11.2 The Existence of Legal Antinomies

11.2.1 The Absence of a Logical Contradiction

In Kelsen's opinion, the legal order is endowed with a logical unity ensured by the existence of the basic norm: "since the basic norm is the reason for the validity of all norms belonging to the same legal order, the basic norm constitutes the unity of the multiplicity of these norms" [12, p. 205]. However, Kelsen admits that "legal organs may create conflicting norms". According to him, a conflict of norms exists "if one norm prescribes a certain behavior, and another norm prescribes another behavior incompatible with the first" [12, p. 205].

Does Kelsen's account mean that we can speak of contradictory rules? Wright notes that "calling two propositions mutually contradictory normally means that they cannot both be true, and calling a set of propositions consistent means that they may, all of them, be true (together)" [28, p. 271]. But norms, since they do not describe a state of affairs but instead prescribe behaviour, "have no truth-value". As Kelsen says, "[a] norm is neither true nor false, but either valid or invalid" [12, p. 205]. This kind of considerations is what leads some authors to consider legal antinomies as things that cannot be assimilated to logical contradictions, i.e., a proposition that states that something is both the case and not the case. Thus, Perelman [22, p. 393] thinks that "antinomies, in so far as they are a concern to law, do not consist in identifying a contradiction, the result of the simultaneous assertion of the truth of a proposition and of its negation, but rather consist in the existence of an incompatibility between directives related to the same object".[5]

On the other hand, Kelsen notices that the assertions describing a normative order may very well be considered to be true or false. This is especially the case for those legal propositions that describe what is or not stipulated in a given legal order. To Kelsen, "logical principles in general, and the Principle of the Exclusion of Contradictions in particular, are applicable to rules of law describing legal norms and therefore indirectly also to legal norms" [12, p. 206]. He thus concludes that "it is by no means absurd to say that two legal norms 'contradict' each other". Nevertheless, even Kelsen [13, p. 214] speaks later of an "opposition" or "antithesis" to underline the distinction between legal antinomies and logical contradictions.

---

[5]Original text: "Les antinomies, dans la mesure où elles concernent le droit, ne consistent pas dans la constatation d'une contradiction, résultat de l'affirmation simultanée de la vérité d'une proposition et de sa négation, mais dans l'existence d'une incompatibilité entre les directives relatives à un même objet".

11.2.2 The Antagonism Between Two Valid Norms

Thus, the antinomy does not challenge the validity of the conflicting norms, since as we have noted validity is not the same as truth. "We cannot claim", Kelsen notes, "that if one of the conflicting norms is valid, the other must be invalid as we will do when it concerns a logical contradiction in which, if a statement is true, the other one must necessarily be false. When we have a conflict of norms, both norms are valid; otherwise, there would be no conflict. Neither of the conflicting norms repeals the validity of the other" [13, p. 213]. In order for legal antinomies to exist, as Perelman phrases it, "two incompatible directives must be prescribed simultaneously and in a way that is equally valid to settle a single situation" [21, p. 399].[6] We have a normative conflict when in a legal system two valid norms coexist, which prescribe one thing and its opposite, and thus undermine the consistency of the legal corpus.

11.2.3 The Pathological Nature of Conflicts Between Norms

According to von Wright, inconsistency and contradiction of norms may be brought to light if we consider the "rationality of demanding and allowing certain things in conjunction with one another" [28, p. 272], because "a norm-giver who demands that one and the same state of affairs both be and not be the case cannot have his demand satisfied. He is 'crying for the moon'. His issuing the norms is irrational" [28, p. 271]. To Kelsen, "to say that a ought to be and at the same time ought not to be is just as meaningless as to say that a is and at the same time that it is not. A conflict of norms is just as meaningless as a logical contradiction" [12, p. 206]. However, he later changes his position, in his *General theory of norms*, as he held that "it is not the case (...) that a conflict of norms (...) makes no sense and that both norms are therefore invalid. Each of the two general norms makes sense and both are valid" [13, p. 214]. If we conclude that "a conflict of norms is meaningless", the author continues, it is "because of the erroneous assumption that two conflicting norms represent a logical contradiction" [13, p. 218].

Nevertheless, legal antinomies appear as failures on the part of the lawmaker, and lawyers try to deal with or eradicate them as a pathological phenomenon. Today, they are branded as a symptom of the deterioration in the quality of the norm-producing process. In addition, the risk of witnessing the emergence of antinomic

---

[6]Original text: "Deux directives incompatibles (doivent être) prescrites simultanément, et d'une façon également valable, pour régler une même situation". Perelman defines legal antinomies as "the existence of an incompatibility between instructions concerning the same object" ("l'existence d'une incompatibilité entre les directives relatives à un même objet") [22, p. 393].

norms is all the more as lawmaking is copious, fragmented, and scattered. Indeed, the "absurdity" sometimes lies at the heart of the lawmaking process and reaches its climax when parallel lawmaking processes bring forward the adoption of similar measures (which, if they are integrated into different corpuses, may produce, in the end, diverging interpretations)[7] or, to the opposite, bring forward contradictory measures, some of them being abrogated even before they come into effect.

11.2.4 The Prevention of Conflicts of Norms

Conflicts between norms appear as failures of the norm producing process that we must try to prevent and eradicate to restore the harmony of the law. The development of the science of legislative drafting is an answer to the will to form a legal corpus that is both consistent and efficient. Conflicts between norms constitute the privileged targets of this process of rationalizing the law that answers Bentham's ambition to "make the ideal qualities of the law more present" (cf. [7, pp. 9 sqq.; 20 sqq.]).[8]

In order to prevent antinomies, one may rely on different strategies. Guides of legislative drafting give writing instructions, whose aim is in the first place to ensure a good articulation between norms. The French *Guide de légistique* develops a rational method of drafting a norm. In particular, it deals with phrases or adverbs that favour the articulation between texts such as "notwithstanding", "by dispensation", "however" (cf. Cards 3.3.2 Choix des termes et des locutions juridiques). A good use of these adverbs prevents possible conflicts between norms by giving more precision to the respective application fields of dispositions. The norm writer, consequently, must make sure the new disposition is well inserted into the legal corpus and, to that effect, he has to take care of the coordination between the norm he is drafting on the one hand, and both present and future texts on the other.

The aims of harmonizing and rationalizing law are widely ensured by abrogation processes that are developed within the frame of a move toward the clarification and simplification of the law developed in France. Impact studies and evaluating processes of norms *ex post* are also methods developed by material science of legislative drafting and, as such, capable of detecting and neutralizing antinomic rules.[9]

---

[7]Cf. the example given by [19]. On the practice of doublets and the risk of giving rise to conflicts between norms, cf. [10].

[8]Original text: "Rendre plus présentes les qualités idéales de la loi. On the reservations that may be inspired by the assumptions of this legal science approach, cf. Millard ([16: 117 sqq.] and especially [16: 125 sqq.]).

[9]On these different approaches, see [6].

11.2.5 The Persisting Presence of Antinomies

In spite of these efforts to rationalize the lawmaking process and notwithstanding the promotion of methods to connect norms, the presence of antinomies persists and seems irreducible – and this for different reasons.

The intractable character of these antinomies is, first of all, related to the heterogeneity of the values that inspired the norms. As Champail-Desplats remarks, "beyond the formal unity under which those [legal] systems find shelter and establish themselves, the material plurality of the values that inspire them and inform them inevitably comes to the fore" [5, p. 61].[10] Legal antinomies therefore reflect the tension among values in our society.

But the problem raised by legal antinomies is also made more acute by how contemporary legislation is evolving. The diversification of norm corpuses and the sectorial compartmentalization of the redaction process make it difficult to coordinate the text writing task and to organize an efficient articulation between normative devices. Thus, even though the wide codifying movement of existing law that began in France in 1989 aimed at keeping the normative inflation in check, while at the same time safeguarding legal consistency, the upshot was a proliferation of conflicts between equally-weighted norms. The reason for this was that the intervention fields of the corpuses was not always clearly defined[11] and the articulation between codified dispositions insufficient. If it is difficult to ensure that normative devices are coordinated, this is due to the interconnection between norms – an interconnection that is admittedly inherent to their integration in a legal system (cf. [21]). This problem, moreover, is intensified because of the greater use of cross-reference mechanisms between texts[12] and the overlapping of the application fields of the corpuses that harbour them.

Another reason for the proliferation of antinomic rules lies in the evolution of the nature of the formulated norms. The increasing complexity, the technical nature, and the specialization of texts makes it more delicate to establish a consistency between them. Hacquet notes: "As it finds its way into a proliferating legal corpus, the law can no longer be a clear and simple text answering a general problematic and creating an overall legislative device. It is now a text that gathers a set of very technical articles modifying previous dispositions or transposing superior norms" [11, pp. 1986 sqq.].[13] Thus, it becomes impervious to the preservation of a harmony

---

[10]Original text : "Par delà l'unité formelle derrière laquelle ces systèmes s'abritent et s'imposent, apparaît inévitablement la pluralité matérielle des valeurs qui les inspirent et les composent".

[11]About the difficult conciliation between French texts concerning usury, distributed both in the Consumption Code and in the Monetary and Financial Code, see Ferrier [8, pp. 219 sqq.].

[12]The cross-reference technique is understood as a formal invitation, stipulated by the rule, to refer to one or several dispositions coming under the same corpus (same code, same law, same decree) or external to it (cf. [17, pp. 55 sqq.]).

[13]Original text: "S'inscrivant dans un corpus juridique foisonnant, la loi ne peut plus être un texte clair et simple, répondant à une problématique générale et posant un dispositif législatif

within this normative muddle, a patchwork of specialized and technical norms that among themselves are not necessarily consistent.[14] Any new disposition that adds, modifies, or subtracts something may have repercussions for existing norms in the same corpus or in an exterior one, a code, a law or a regulation. Two main difficulties jump here to the fore: on the one hand, how to coordinate norms depending on the more general law with those depending on specific laws; on the other hand, how to coordinate specialized norms among themselves.

If the existence of legal antinomies is widely acknowledged, the question about how to handle them is the topic of an ongoing discussion. Within this debate, some authors suggest a categorization for different kinds of conflicts, because they think that it is necessary to propose ways of dealing with them adapted to their specific nature.

11.3   Different Types of Antinomies and Different Ways of Dealing with Them

11.3.1   Deontic Antinomies and Axio-teleological Antinomies

A distinction has been proposed between deontic and axio-teleological antinomies. Deontic antinomies "can be observed *in abstracto* just when reading the structure of the statements ('it is permitted to smoke' versus 'it is forbidden to smoke')" [5, p. 61][15] and should be differentiated from axio-teleological antinomies, which "arise *in concreto* when a decision has to be made" (5, p. 62).[16] Whereas deontic antinomies can be observed directly in the wording of conflicting norms, axio-teleological antinomies are only revealed by the interpretation that has to occur when the norms are implemented. Thus, it is the interpretation of statements that is responsible for the emergence of the axio-teleological conflict: through their interpretation, these statements "prove to be contradictory axiologically or teleologically" (5, p. 60).[17] For instance, the freedom of expression may, in some legal cases, conflict with the right to respect for private life.

As suggested above, the difference between kinds of antinomies is strictly linked to different ways of addressing them: whereas deontic antinomies might be solved

---

d'ensemble. C'est désormais un texte qui regroupe un ensemble d'articles très techniques qui modifient des dispositions antérieures ou transposent des normes supérieures".

[14]On the difficulties as to how to harmonize the status of protected employees because of the diversity of mandates – which themselves were a result of the codification of the French Labor Code into existing law, cf. [15, p. 842].

[15]Original text: "S'observent *in abstracto* à partir de la seule lecture de la structure des énoncés ('il est permis de fumer' versus 'il est interdit de fumer')".

[16]Original text: "Emergent *in concreto* à l'occasion d'une décision à prendre".

[17]Original text: "S'avèrent axiologiquement ou téléologiquement contradictoires".

through the application of logical principles of conflict resolution (hierarchical, chronological, specificity principles), these meta-norms of interpretation would not be appropriate to bring about an answer to axio-teleological antinomies (5, p. 62). The reason for this is that teleological antinomies do not provide any temporal differentiation, hierarchy, or difference in the degrees of generality.[18]

However, several remarks can be made on this distinction. First of all, identifying a conflict of norms always implies an act of interpretation. Thus, Perelman notes "in law, a purely formal, i.e. literal, contradiction is not sufficient to give rise to an antinomy since, when the judge interprets texts, he can give to the same terms a different meaning or a different application field in order to avoid a conflict between norms; he can also dismiss the application of one norm either because it is in opposition with a superior law or because he considers it has been tacitly abrogated by a posterior law" [22, p. 404].[19] "An antinomy is never purely formal", the author continues, "because the understanding of a legal rule implies its interpretation".[20] And indeed, it is when texts are being implemented that a conflict between norms will be either revealed or sidestepped.

Moreover, behind a deontic antinomy, there often looms in the background an axio-teleological antinomy. Thus, in order to deal with the deontic antinomy it is often necessary to take into account the underlying conflict of values. For instance, to resolve the contradiction between the norms "it is forbidden to smoke"/"it is permitted to smoke", the conflict between two values will have to be resolved first: the value of individual freedom, which inspires the latter rule and the value of the protection of public health, on which the former is grounded (comp. 5, p. 67).

Accordingly, Perelman suggests we resolve antinomies by looking for the "justification" of the rules, i.e. the more general principles that justify the conflicting rules. Therefore, the judge should look for the foundation of each of the conflicting norms and then weigh the values or interests protected by antinomic norms.

This approach will be challenged by Bobbio, who notes that "our legal systems are not unified ethical ones, that is to say they are not based on a single ethical postulate or on a group of consistent postulates but they are based on several values which are often antinomic" [3, p. 91].[21] Bobbio [3, p. 241] continues: "How

---

[18]With respect to the inadequacy of these interpretation principles to solve the antinomies between human rights (cf. [26]).

[19]Original text: "En droit une contradiction purement formelle, c'est à dire littérale, ne suffit pas pour donner lieu à une antinomie, car le juge, en interprétant les textes, peut donner aux mêmes signes un sens différent ou un autre champ d'application de façon à éviter le conflit de normes; il peut aussi écarter l'application de l'une des normes, soit parce qu'elle s'oppose à une loi supérieure, soit parce qu'il la considère comme tacitement abrogée par une loi postérieure".

[20]Original text: "L'antinomie n'est jamais purement formelle, poursuit l'auteur, car toute compréhension d'une règle juridique implique son interprétation".

[21]Original text: "Nos systèmes juridiques ne sont pas des systèmes éthiques unitaires, c'est à dire, ils ne se fondent pas sur un unique postulat éthique, ou sur un groupe de postulats cohérents, mais ils sont des systèmes à plusieurs valeurs et ces valeurs sont souvent antinomiques entre elles".

could we pretend we have resolved an antinomy between rules by means of the justification device when this very device may bring about the discovery of an antinomy between values and, consequently, bring about the possibility of justifying two rules, depending on whether we refer to one value or the other?"[22] By resorting to the justification of rules, Bobbio is afraid that we are made hostage of the personal assessments of the interpreter to solve the conflict of values.

Whereas the man in the street would, among two conflicting rules, favour "the fairer one", the law, according to Bobbio, generally gives "a different answer and gives the interpreter some criteria to help him choose without expressing personal preferences" [3, p. 241].[23] Bobbio thinks that, as far as the resolution of antinomies is concerned "the choice between two conflicting rules is not primarily left to the judge but is made according to traditional criteria of preference between rules, criteria that forbid, except in extreme cases (. . .), a discretionary decision on the interpreter's part" [3, p. 239].[24] The judge should then find a solution to legal antinomies through resorting to these "traditional criteria" that are the interpretation of norms. These logical principles would then permit them to provide a non-arbitrary resolution of the conflict.

11.3.2 The Meta-norms of Interpretation

The logical interpretation principles constitute "rules of the art" ("Règles de l'art") [2, p. 9] shaped by legal theory and judicial practice, and are what makes the legal corpus' consistency possible. As Kelsen says: "since the cognition of law, like any cognition, seeks to understand its subject as a meaningful whole and to describe it in non-contradictory statements, it starts from the assumption that conflicts of norms within the normative order which is the object of this cognition can and must be solved by interpretation" [12, p. 206]. The judge should then resort to rules of interpretation to solve the conflict and determine the major premise of the judicial syllogism.

It is in fact, as Troper [27] notes, a twofold syllogism that is implemented by the judge. Before implementing the primary syllogism in which the major premise, the minor premise, and the inferred conclusion are constituted, respectively, by the

---

[22]Original text: "[. . .] comment on pourrait prétendre résoudre une antinomie entre des règles au moyen du procédé de justification, lorsque ce même procédé peut conduire à la découverte d'une antinomie de valeurs et, par conséquent, à la possibilité de justifier les deux règles, selon que l'on s'en rapporte à l'une ou à l'autre valeur".

[23]Original text: "Le choix entre deux règles incompatibles n'est préliminairement pas confié au juge, mais est réglé par des critères traditionnels de préférence entre les règles, critères qui, sauf en des cas extrêmes [. . .], interdisent une décision discrétionnaire de l'interprète".

[24]Original text: "Le choix entre deux règles incompatibles n'est préliminairement pas confié au juge, mais est réglé par des critères traditionnels de préférence entre les règles, critères qui, sauf en des cas extrêmes [. . .], interdisent une décision discrétionnaire de l'interprète".

relevant legal rule, the case submitted to the judge, and the jurisdictional decision made by the judge, another syllogism must previously take place to determine which legal rule is to be applied. This preliminary syllogism, whose major premise is an interpretation rule and the minor one the applicable legal rule or rules, is what enables the judge to determine, thanks to its conclusion, which is the major premise of the primary syllogism.

At the level of the preliminary syllogism, three principles are usually invoked to interpret norms so as to avoid conflicts between them: *lex superior derogat inferiori* (hierarchical criterion), *lex posterior derogat legi priori* (chronological criterion), *lex specialis derogat generali* (specificity criterion).

The first principle is used to resolve vertical conflicts between norms that, according to Kelsen's model, are located on different levels of the pyramid of norms. This criterion is often brought to bear in the motivation of justice decisions whenever the conformity of regulations to a law, or the conformity of a law to international treaties or to constitutional dispositions is at stake. For instance, the introduction in the French system of the Priority question of constitutionality in front of the judicial and administrative judge has provided a wide range of applications to this meta-norm in the jurisdictional debate.

The other two criteria, on the other hand, are rarely invoked explicitly in the motivations of judgments. Kelsen, incidentally, does not consider the chronological criterion as an implement meant to resolve antinomies. He denies the existence of a conflict "between a norm and the abrogating one suppressing its validity, because the first norm stops being valid when the second one comes into validity" [13, p. 213]. Yet the conflict between norms implies a contradiction between valid norms. However, the abrogation is often only implicit and requires an interpretation of new and old norms to investigate whether the new one indeed abrogates the previous one, or if it only constitutes a specific norm infringing the old general norm. The judge will then often have to combine the application of chronological criteria with criteria of specificity.

Beyond the question of the legal value of these meta-norms, which is still being discussed, one may raise the question of their practical efficiency. If they prove themselves not very effective in dealing with conflicts between values, one may challenge the importance ascribed to them in the resolution of deontic conflicts. This is especially the case if one considers that a conflict between values often underlies the deontic conflict – something that we already had the occasion to notice. Thus, the meta-norms cannot really address this kind of conflict, except, of course, if we propose a hierarchy between values. And clearly the judge will not always feel disposed to ignore the conflict between values underlying the conflict between norms. Instead, he will be tempted to resort to other techniques of interpretation to resolve the antinomies.

11.3.3 The Diversity of Methods of Interpretation

The influence of these principles of logical interpretation is, in fact, diminished by the wide diversity of interpretation techniques the judge has at his disposal and the wide freedom of interpretation they entail (see [2, p. 41]).

So, in order to deal with conflicts between rights of equal value, the European Court of Human Rights has developed a method of weighing the present interests. This method is also followed in France by the *Cour de Cassation* when, for instance, a demand of repair of an infringement in respect for private life is submitted. In these cases, the Court has to assess whether the applied sanction constitutes an acceptable intrusion of civil responsibility in the liberty of the press. The court, moreover, has to check whether the judge has set the damages of this infringement to the respect for privacy "according to a reasonable, proportional relation between the imposed sanction and the rightful target aimed by it".[25] The right to the respect for private life and the liberty of the press are weighed against each other; the protection of the former must not entail a disproportionate infringement of the latter.

Given this diversity of interpretation methods, the judge has considerable freedom to simply recur to the most appropriate technique to obtain the solution he deems most suitable. This is what has been addressed by Carbonnier [4, n° 160] as "tactical eclecticism" – an eclecticism that may go as far as to negate the very presence of conflict.

11.3.4 The Negation of the Antinomy

Very often, the judge's pragmatism will bring him to deny the presence of an antinomy by ensuring the conciliation between potentially conflicting norms. As several writers such as Ost [20, pp. 7 sqq.; 163 sqq.] have noted, judges tend to interpret applicable norms so as to make them compatible. They delimit the respective fields of application of the two conflicting norms in order to eradicate the antinomy. Conciliation is an "executory art" ("art d'exécution"), which consists in conciliating diverging interests, a "mediation competence" ("compétence d'arbitrage"), whose aim is to articulate the application of antinomic norms (see [26]). Von Wright wrote: "I assume that in practice the conflict is often removed by some 'modifications' in the conflicting norms, restricting their content so that the contradiction is eliminated" [28, p. 278].

We can take the example of the antinomy that exists between article 2052 of the French Civil Code and previous article 888 as well as the new article 890 of the same Code. Whereas the first text forbids the rescissory action concerning a transaction,

---

[25] Original text: "Dans un rapport raisonnable de proportionnalité entre la sanction imposée et le but légitime visé" (Civ. 1re, 21 févr. 2006, Bull. civ. I, n° 97; RJPF mai 2006. p. 12, note Putman. – Cf. CEDH 22 oct. 2007, n° 21279–02, D. 2007. 2737, obs. Lavric (4)).

the other two permit it for acts proceeding to the division of a joint ownership. The previous article 888 states that a "rescissory action is admitted against any act whose aim is to terminate the joint ownership among co-heirs, no matter whether it is called a sale, an exchange, a transaction or any other name".[26] On the other hand, according to the wording of new article 890, "[t]he action for complementary share is admitted against any act, independently from its denomination, whose aim is to terminate the joint ownership among those who share it".[27] When the agreement that terminates the joint ownership has the characteristics of a transaction, the *French Cour de Cassation* favours the dispositions that govern the partition. Thus, the Court judged that "the agreement whose aim is to terminate the joint ownership among spouses is susceptible to a rescissory action even if it contains mutual concessions by the parties and so constitutes a transaction".[28] The conflict between norms is thereby solved through delimiting the field of application of the two texts.

The negation of the antinomy can also be the result of the negation of one of the two conflicting norms. Thus, the articles 2048 and 2049 of the French Civil Code provide diverging interpretation instructions as far as transactions are concerned. The first text underlines the fact that "the transactions are limited to their object: renouncing by means of a transaction to all the rights, suits, and claims concerns only what is relative to the conflict that gave rise to it".[29] However, in the second text, we find a specification that states that "a transaction will only solve the conflicts addressed by it, no matter whether the parties expressed their intention through general or specific expressions or whether this intention can be recognized as a necessary result of what was expressed".[30] The article 2048 and the beginning of the article 2049 provide for a strict interpretation of the transaction whereas article 2049 *in fine* invites us to a more flexible interpretation. This antinomy is the transposition in the French Civil Code of the diverging opinions of two authors who inspired the codifiers dealing with transactions: whereas article 2048 is inspired by Ulpian, article 2049 *in fine* refers back to Domat's positions (cf. [9, p. 390]).

---

[26]Original text: "L'action en rescision est admise contre tout acte qui a pour objet de faire cesser l'indivision entre cohéritiers, qu'il fut qualifié de vente, d'échange et de transaction, ou de toute autre manière".

[27]Original text: "L'action en complément de part est admise contre tout acte, quelle que soit sa dénomination, dont l'objet est de faire cesser l'indivision entre copartageants".

[28]Original text: "La convention ayant pour objet de faire cesser l'indivision entre les époux est sujette à l'action en rescision même si elle comporte des concessions réciproques entre les parties et constitue une transaction" (Civ. 1re, 9 janv. 2008, pourvoi n° 06–16.454, LPA 9 avr. 2008, p. 8, note Ph. Malaurie; RTD civ. 2008 p. 342, obs. M. Grimaldi).

[29]Original text: "Les transactions se renferment dans leur objet : la renonciation qui y est faite à tous droits, actions et prétentions, ne s'entend que de ce qui est relatif au différend qui y a donné lieu".

[30]Original text: "Les transactions ne règlent que les différends qui s'y trouvent compris, soit que les parties aient manifesté leur intention par des expressions spéciales ou générales, soit que l'on reconnaisse cette intention par une suite nécessaire de ce qui est exprimé".

Confronted with these diverging interpretative instructions, case law tends to favour a narrow idea of the object of the transaction but it sometimes takes advantage of this normative ambiguity to act as if entitled to propose a flexible interpretation of the transaction agreement. The antinomic norms would thus both preserve their vocation to be applied to the factual situation, and the judge would maintain the liberty to choose, for each case, the applicable norm, thereby negating the existence of the other one.

11.4 Conclusion

The existence of an antinomy challenges the applicability of the conflicting norms. One of them will have to be applied to the detriment of the other or each of their application fields will have to be defined in order to make them applicable simultaneously. It is up to the judge to deal with the antinomy in the most appropriate manner; after identifying the conflict, he will choose how to proceed. The options at his disposal are the following: negation, conciliation, weighing the present interests, and applying logical interpretation principles. The legal antinomy phenomenon invites us to dismiss our belief in a rationality that would be inherent to the law and to admit that case law has to play a part of the guardian's role concerning legal consistency. It is a secondary consistency – which is put into music by the judge, and sometimes not without dissonance.

References

1. P. Amselek, *Cheminements philosophiques dans le monde du droit et des règles en général* (Armand Colin, Le temps des idées, Paris, 2012)
2. Y. Aquila, Cinq questions sur l'interprétation constitutionnelle. Revue française de droit constitutionnel **21**, 9–46 (1995)
3. N. Bobbio, Des critères pour résoudre les antinomies. Dialectica **18**, 237–258 (1964)
4. J. Carbonnier, *Droit Civil I* (PUF, collection Quadrige, Paris, 2004)
5. V. Champeil-Desplats, Raisonnement juridique et pluralité des valeurs: les conflits axio-téléologiques de normes, in *Analisi e diritto*, ed. by P. Comanducci, R. Guastini (Giappichelli, Torino, 2001), 59–70
6. S. Chassagnard-Pinet, L'appréhension des conflits de normes de même niveau par la légistique. Vers une prévention et un traitement méthodique des antinomies du droit ? in *Les conflits de normes*, ed. by F. Péraldi Leneuf, S. Schiller. Le traitement légistique et jurisprudentiel des conflits horizontaux de normes (GIP Mission de recherche Droit et Justice, Paris, 2012), 27–50
7. J.-P. Duprat, Genèse et développement de la légistique, in *La confection de la loi*, ed. by R. Drago (PUF, Cahiers des sciences morales et politiques, Paris, 2005), 9–45
8. N. Ferrier, Les incertitudes du régime de l'usure liées à sa codification. Contribution à l'analyse critique de la « codification-compilation. Rev. Trim. Droit Com. **2**, 219–242 (2005)
9. P.-Y. Gautier, Où, à l'occasion de l'examen de la portée d'une transaction, la Cour de cassation prend en faute Domat et fait, grâce à lui, produire un effet utile aux règles d'interprétation qui y président. Rev Trimestrelle de Droit Civil. **2**, 390–394 (1995)

10. C. Goldie-Genicon, *Contribution à l'étude des rapports entre le droit commun et le droit spécial des contrats* (LGDJ, Paris, 2009)
11. A. Haquet, Les études d'impact des projets de loi: espérances, scepticisme et compromis. Actualités Juridiques – Droit Administratif **36**, 1986–1993 (2009)
12. H. Kelsen, *Pure Theory of Law* (University of California Press, Berkeley/Los Angeles, 1978) (first edition, Franz Deuticke, Vienna 1934)
13. H. Kelsen, *General Theory of Norms* (Clarendon, Oxford, 1991)
14. Ph. Malaurie, *Les antinomies des règles et de leurs fondements,* in *Mélanges Pierre Catala* (Litec, Paris, 2001), 25–31
15. C. Maugüé, A. Courrèges, La méthode et le processus de recodification. Actualités Juridiques – Droit Administratif **16**, 842–850 (2008)
16. E. Millard, Les limites des guides de légistique: l'exemple du droit français, in *Guider les Parlements et les gouvernements pour mieux légiférer. Le rôle des guides de légistique*, ed. by A. Flückiger, C. Guy-Ecabert (Schulthess, Zürich, 2008), 117–128
17. N. Molfessis, Le renvoi d'un texte à un autre, in *Les mots de la loi*, ed. by N. Molfessis (Economica, Paris, 1999), 55–72
18. Montesquieu. *The Spirit of the Law* (translated by A.M. Cohler, B.C. Miller and H. Stone) (Cambridge University Press, Cambridge, 1989)
19. H. Moysan, L'humour n'est pas exclusif du sérieux (quand la confection de la loi se passe d'un examen préalable approfondi de l'état du droit). La semaine juridique. Administration et Collectivités territoriales **29**, 3 (2010)
20. F. Ost, L'interprétation logique et systématique et le postulat de la rationalité du législateur, in *L'interprétation en droit, Approche pluridisciplinaire*, ed. by M. van de Kerchove (Publications des Facultés universitaires de Saint Louis, Bruxelles, 1978), 97–184
21. F. Ost, M. van de Kerchove, *Le système juridique entre ordre et désordre* (PUF, Paris, 1988)
22. C. Perelman, Les antinomies en droit. Essai de synthèse. Dialectica **18**(1–4), 391–404 (1965)
23. C. Perelman, L'interprétation juridique. Archives de philosophies du droit **17**, 29–37 (1972)
24. J.E.M. Portalis, Discours préliminaire du premier projet de Code Civil (1801), in *Écrits et discours juridiques et politiques,* ed. by J.E.M. Portalis (Presses Universitaires d'Aix-Marseille, Aix-en-Provence, 1988), 21–34
25. P. Rémy, Eloge de l'exégèse. *Revue de la recherche juridique* **2**, 254–262 (1982)
26. V. Saint James, Hiérarchie et conciliation des droits de l'homme, in *Dictionnaire des droits de l'homme*, ed. by J. Andriantsimbazovina, H. Gaudin, J.-P. Marguenaud, S. Rials, F. Sudre (PUF, Paris, 2008), 378–381
27. M. Troper, Fonction juridictionnelle ou pouvoir judiciaire? Revue Pouvoirs **16**, 5–16 (1981)
28. G.H. Von Wright, Is there a logic of norms? Ratio Juris **4**(3), 265–283 (1991)

Chapter 12
The Service Contract (Contrat d'Entreprise)

Problems and Methods of Classification

Juliette Sénéchal

Abstract The service contract, on the basis of the general definition given by the Civil Code (Art. 1710), is defined by the French case law as a contract whereby a person called the contractor undertakes the obligation to carry out a defined task on behalf of a principal in view of a remuneration, and without being either an agent or an employee of the principal. The economic importance of the service contract is at present unfortunately not reflected by a clearly determined legal identity or a strong and consistent statutory framework. These shortcomings call for a proposal of reform. We may choose between proposing either a reform of the service contract law as it appears in the existing civil law, or a solution that is part of a wider reform of the law of special contracts. These two options are each underpinned by a different vision of the legal system. Indeed, proposing a reform of service contract law inside the existing positive law amounts to giving priority to the typological vision of contractual types, a vision marked by redundancies and shortcomings. We will follow instead the spirit of the Civil Code of 1804, a spirit that is underlain by exigencies of completeness and openness for the legal system, so as to propose a new identity and a new legal framework for the service contract. This in turn is part of a wider reform of the law of special contracts based on classification.

12.1 An Observation on the Shortcomings of Service Contracts

The service contract (in the French Civil Code: *contrat de louage d'ouvrage et d'industrie*, in modern legal language: *contrat d'entreprise*), on the basis of the general definition given by the Civil Code (Art. 1710), is defined by French case law as a contract whereby a person called the contractor undertakes the obligation to carry out defined work on behalf of a principal, in exchange for remuneration, and without being either an agent or an employee of the principal. The service to be

J. Sénéchal (✉)
Equipe René Demogue (EA 4487), Université Lille Nord de France (UDSL), Lille, France
e-mail: juliette.senechal-2@univ-lille2.fr

© Springer International Publishing Switzerland 2015   249
M. Armgardt et al. (eds.), *Past and Present Interactions in Legal Reasoning and Logic*, Logic, Argumentation & Reasoning 7, DOI 10.1007/978-3-319-16021-4_12

performed may be either material (e.g.: cleaning, building, manufacturing, etc.), or intellectual (e.g.: teaching, counseling, drafting, healing, etc.).

The service contract is very important at the economic level, particularly from a European internal market perspective.[1] As Benabent [1, n. 472] states, it "has become the second pillar of an economy based on goods and services: in the domain of services, it holds the same place as sales in the domain of goods". More precisely, the service contract is not only the pillar of the service sector, but also a tool that rivals sales in the goods sector.

Its omnipresence on the economic stage makes it an indispensable tool for many professionals.

This is the case for the medical professions, barristers and counsellors, management consultants, chartered accountants, consultants in data processing and information systems, architects, construction firms and all other manufacturers, middlemen with no representatives, turnkey factory creators, garage owners, plumbers, dry cleaners, cobblers, and automatic car washing companies, amongst others.

Unfortunately, the economic importance of the service contract is at present unaccompanied by a clearly determined legal identity or a strong and consistent statutory framework (cf. [7, 8]).

The 1710th article of the Civil Code, which is inserted into the chapter concerning hiring, proposes a very vague definition of the service contract, further hindered by obsolete terminology: "hiring services is a contract by which one of the parties commits itself to do something for the other, for a price on which they have agreed". Indeed, the term: "hiring services" has not stood the test of time since it has been replaced by "service contract". Such a substitution can easily be justified: the service contract cannot be defined as hiring since it does not possess the characteristic common to all hiring, i.e., the temporary use of something.

Moreover, Article 1710 is not followed by any articles explaining the guidelines that are to be applied to service contracts. Nor has it been possible to correctly remedy these original shortcomings in the service contract through case law, by lawmakers, or by doctrine, in the two centuries that have followed the writing of the French Civil Code in 1804.

---

[1] Green paper from the Commission on policy options for progress towards a European Contract Law for consumers and businesses, COM (2010), 1st July 2010: "*4.3.3. Should specific types of contracts be covered by the instrument?* In addition to general contract law provisions, the instrument could contain specific provisions for the most prevalent types of contract. The most common and relevant from the internal market perspective is the contract for sale of goods. Service contracts are also very important. However, given their heterogeneous character, specific provisions will have to be made for specific types of service contracts. For example, the instrument could contain provisions for 'sale-like' service contracts, such as car lease, or for insurance contracts. Furthermore, contracts in the financial services area are of a very specific and technical nature, particularly when concluded between professionals, and need a prudent approach as the legal environment in these areas changes rapidly". Proposal for a regulation of the European Parliament and of the Council on a common European sales law of the 11 October 2011, COM 2011 (635) final.

The service contract was removed from the category of "hiring" in which it was initially and ill-advisedly placed by the writers of the Code; it has progressively become a type – excluding a contractual species – that is added to other types of contracts specified in the Code. Its inclusion in a typology has not been able to save it from the initial uncertainty of its nature or its almost non-existent statutory framework.

The service contract has remained a heterogeneous notion, unable to benefit from a complete and consistent basis in statute. The perpetually vague nature of the service contract has reinforced the almost complete non-existence of a specific, coherent, and comprehensive statutory framework, despite numerous case law interventions.

These case law interventions have essentially consisted in attributing one of two legal systems to the service contract: either that relating to the law of obligations, or the application of certain rules derived from construction contract law. These ad hoc interventions have, however, proven ineffective. At best, they have shown that a single framework for the service contract is not possible because the contract is too heterogeneous; and that two distinct sets of rules for two distinct subclasses of the service contract would be more appropriate.

However, case law has not yet managed to reveal a clear dividing line within the service contract between two homogeneous subclasses, each one with its accompanying statutory framework. That is why it has been necessary to work with two divisions: first, the division between service contracts creating a principal (primary) obligation of skill and care and those creating a principal (primary) obligation to achieve a result; and second, the division between contracts that involve a material service and those that do not.

The study of positive law concerning service contracts results, then, in a very disappointing observation: the service contract, fundamental at an economic level, is not and cannot be represented in any way or by any set of legal rules that reflect its economic importance, because of its place within the system of special contracts. It is an *ineffective* contract that does not fulfil the function bestowed upon it: to propose consistent and complete regulations for contracts regarding the provision of services for a price. It is, moreover, a redundant contract. Indeed, because of its undetermined and heterogeneous nature, it becomes a "miscellaneous" category, and thus performs the same task as the unnamed contract (*contrat innommé*, Article 1107 of the Civil Code, which is the official contractual category functioning as a point of entry, or gateway, into the system) (see Fig. 12.1).

The service contract, as it was initially defined and regulated by the writers of the Civil Code, and as it has subsequently been understood in case law, has an undetermined and quasi-non-existent statutory framework.

These shortcomings, unacceptable in the face of the economic importance of this contract, call for a proposal of reform, requiring a modification of the boundaries between this contract and neighbouring ones.

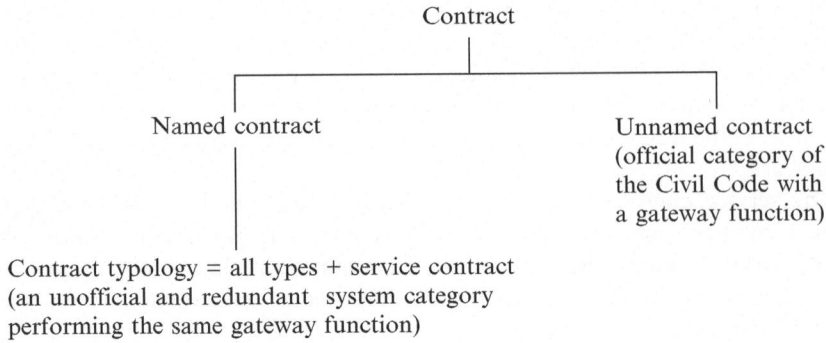

Fig. 12.1 Contract, named contract, unnamed contract, contract typology

12.2   A Preliminary Question of Method: The Necessary Search for the Relationship Between the Whole and the Part, Typology of Classification

We think that the seeming impossibility of salvaging the service contract is due to the existing relationship in positive law between Article 1710 of the Civil Code and the legal system proposed by the Civil Code with regard to contract law.

A methodological consequence results from this observation. Insofar as two main types of relations between the whole and the part coexist, we need first to present these two types of relationships in order to determine in which of them we will find the service contract and all the named or unnamed contracts in the Code. These two general categories of relationships between the whole and the part are classification and typology.

12.2.1   Classification

The first general category of relationship between the whole and the part, classification, constitutes, in terms of formal logic, a perfect system. On the logical level, a system is perfect when, at a given moment, systematized concepts are *collectively exhaustive* as they necessarily include all the units existing at that moment and *mutually exclusive* as a unit is supposed to be part only of a concept with a given hierarchical rank, excluding any other similarly ranked concept.

Classification can be analysed as an organization of abstract concepts, from the more general to the more specific. The classifying system takes, in that respect, the shape of a tree or of a pyramid, depending on whether the most general concept is placed at the top or at the bottom of the hierarchical organization.

The classification technique consists in bringing together all the species of the same genre and giving each of them their own rank within that shared genre. This technique thus possesses a vertical dimension that shows the existence of an "overall relationship". Each term in the hierarchy is a species of a superior genre and the relationship in meaning is one of genre to species and of species to subspecies if the concepts are positioned as a pyramid or, if they are positioned as a tree, a relation of species to superior species. In this type of organization, it is necessary, in order to define the precise nature of a species, to add to its specific quality that of the genre in which it is included.

Classification also possesses a horizontal dimension. On a horizontal level, i.e. within the same hierarchical rank, concepts are "parallel species". Parallel species are concepts that are included in a similar genre and that consequently possess, by definition, a similar number of qualities, attributes, and/or predicates. They therefore all possess a definition that offers the same degree of precision, composed not only of the qualities that belong to their common genre but also of the quality that is unique to each of them. This quality, belonging to a given species, constitutes the difference specific to the species in question and allows us to distinguish it from parallel species.

Owing to the sum of characteristics contained in the notion of classification, a concept that is inserted inside a complete and sophisticated classifying system can have only one 'Aristotelian' definition, also known as a 'real' definition, which is arrived at through considering the closest genre and the specific difference.

This notion of an Aristotelian definition is the formula that expresses the essence of something (cf. *Topics* 102a3; *Met.* 1030a6). This essence consists of the qualities of the genre plus the specific difference, which is the quality that distinguishes one species from others of the same sort. In the Aristotelian definition, the most important thing is the distinction between the meaning, the essence and the nature of the concept to be defined and the meaning, the essence, and the nature of neighbouring concepts.

From the perspective of classification, a subspecies thus needs the definition of its species, that of its genre and that of its parallel subspecies, so that its nature can be completely determined. As an example, the contract as a genre is a dual expression of wills with a view to creating legal consequences. The service contract as a genre is a contract (an implicit reference to the specificity of the genre) by which a contractor commits himself to do something for another party for a price. A subspecies of the service contract is the construction contract (an implicit reference to the specificities of the species and the genre) concerning the construction of a building. The construction contract is in turn different from the transportation contract, the other subspecies of the service contract, in virtue of a specific difference.

12.2.2   Typology

Unlike concept classification, understood as a phenomenon that is first unitary and then subject to subdivision, typology can be analysed as a useful federating word to represent the sum of the elements that compose it.

First, there is a list of concepts (or types) obtained through a process of induction. Subsequently, the adding together of the whole set of these concepts forms a typology. In other words, the type precedes the typology in contrast to classification, which precedes the species.

Legal types, which precede typology, also come with a terminological definition.

In opposition to the Aristotelian definition that is obtained through "closest genre and specific difference," the terminological definition includes all the qualities specific to the notion. *Any overlap with a neighbouring contractual category, as well as overlaps with the overall system, is without importance. Both redundancy and gaps in the definition are permitted.*

A typological reading of the definition of the service contract might be as follows: "the expression of two wills with a view to carrying out an economic operation by which one of the contractors does something for the other for a price". All the qualities of the contractual type are included in its definition without the necessity of referring to a more general category (such as the contract) a more specific category (such as a building contract) or a parallel category (such as the hiring of things).

As we have just stated, typology comes only after the types which compose it. It is a list, a sum, of the types we have arrived at through induction. It can nonetheless be formalized in a continuous series of concepts on a horizontal line in which each point denotes a concept. In this list, most of the concepts will be equally general or specific. Almost all the types falling under a typology will have the same number of qualities and attributes. In that respect, it should also be said that a type is often very precise. Generally, it has many qualities, and many attributes. This characteristic marks another difference from the notion of classification since classification has, in addition to a horizontal dimension, a vertical one. It is broken down into genre, species, and subspecies that by their very nature do not have the same degree of determination.

This typological list may however be incomplete. It is almost certain that the induced types are not and will never be able to form an exhaustive list. Indeed the non-generalization of the inductive qualities results in the appearance of very specific types, incapable of covering the whole field of all present and future realities. This incompleteness also marks a difference with the notion of classification, which must essentially tend toward exhaustiveness.

In this typological list, there may also be overlaps or redundancies. The categories, identified through a process of induction, may in reality represent social phenomena which could fall under several induced categories. But no overlaps are possible in the classification method.

Finally, one type may have more characteristics than another, thus implying that it has a more precise description. It is, however, inadmissible to establish a genre-to-species relationship between two types. The type to sub-type relationship does not have the same legal impact as the genre to species relation. Indeed, a sub-type is in no way supposed to receive a quality belonging to a more general type in order to complete its definition.

12.3   Implementing the Chosen Method to Make Up for the Shortcomings of the Service Contract: Classification

To make up for the shortcomings in the current provision of service contracts, we may propose either a reform of service contract law as it is contained in existing civil law, or a solution that is part of a wider reform of the law of special contracts. Each alternative is underpinned by a different vision of the legal system.

Indeed, proposing a reform of the law of service contracts from within existing positive law amounts to giving priority to the typological vision accepted by the modern lawmaker: a pragmatic vision of contractual types which tolerates the existence of a typology full of redundancies and shortcomings.

However, we choose to follow the spirit of the Civil Code, a spirit that is underlain by exigencies of completeness and openness within the legal system. We will put forward a way of making up for the shortcomings in contract law regarding the provision of services, which is itself part of a wider reform of the law of special contracts.

We have thus undertaken to propose a law concerning provision of services for a price, by framing a contract that is part of a contractual classification respecting logical constraints [2]. (a) This method leads us to suggest three levels of classification for the service contract: a genre, the contract concerning isolated and definite provision; two parallel species, immediate or deferred sale; and three subspecies, the activity agreement (in French: *contrat d'activité*), the 'piece of work' contract (in French: *contrat d'oeuvre*), a service contract in which the contractor controls all the elements of the action and the 'work' contract (in French: *contrat d'ouvrage*), in which some of the elements are imposed alongside an intrinsic definition based on a positive quality, the notion of specific work. (b) This method will then allow us to specify the respective statutes regulating the service contract and its three subspecies.

12.3.1   The Choice of a Legal System Based on the Technique of Classification and Respecting Logical Constraints

As just stated, in order to propose a way to make up for the current shortcomings of the service contract, we have chosen to preserve the spirit of the Civil Code, a spirit underlain by exigencies of completeness and openness in the legal system.

This choice thus supposes inserting the changes in service contract law into a wider reform of the law of special contracts.

We have chosen to get rid of the typological aspects of the organization of contracts as set out in the Civil Code and to keep only the classifying aspects. Indeed, we hold that classification alone is really able to do justice to the spirit of the comprehensive and transparent Civil Code.

This vision of the legal system and the framing of our proposal to reform service contract law within a wider reform of contract law are necessary, because we are convinced that adequate regulation of the provision of service contracts can be brought about only by changing the boundaries between the service contract and those contracts which resemble it closely. More precisely, this is to be done by redefining the respective domains of contract of sale and service contract.

Our choice of method, borrowing from the spirit of the French Civil Code (which treats the law as a comprehensive and open system), is a result of the spirit of this Code being the most apt to provide a consistent basis of civil law rules for the legal Europe currently under construction.

The open character of the system of classification must be served by concepts with a very open definition allowing for the inclusion of a maximum number of concrete realities. The comprehensive character of the system must be served by a clear and complete pyramid of genre and species. Even though this technique is difficult, it is the best fitted to satisfy the all-encompassing, academic, and rationalist spirit of the system. Moreover, the classification and organization of concepts offers a richer treatment than typology and so may be easily reworked into a typology in the future, whereas the inverse is not possible.

Furthermore, this system of classification is subject to a logical constraint that cannot be ignored: the respect of the rules set out by Blanché [2]. According to these rules, at the same hierarchical rank within a classification, it is possible to obtain concepts with a determinate nature and which cover the whole domain of the genre in which they are included, but only as a triad of parallel species: two extreme opposites and an intermediary position. That is to say, two extreme species and one hybrid. Between black and white, there is always room for grey (see Fig. 12.2).

Yet the categories that appear in the system will not all be determinate and homogenous. Thus, as in the Civil Code, we reconcile the system's complete character with its open one by inserting into the system one or more contractual dyads like the one formed by specified and unspecified contracts. These fundamental dyads allow us to insert openings into the legal system toward concrete realities that have not yet been sanctioned by the legal order. This category of dyad consists of one restrictive and homogeneous category and one that is open and heterogeneous. Thus, in the dyad of specified and unspecified contracts, the specified contract

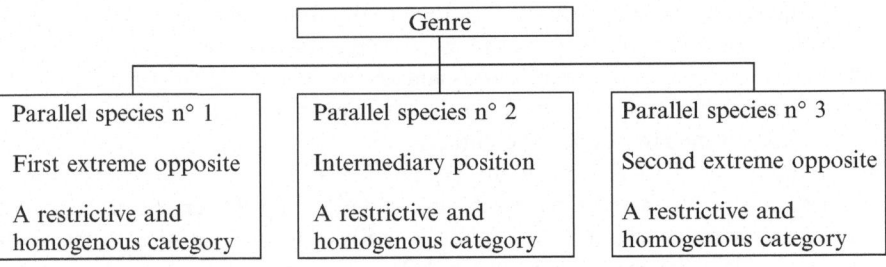

Fig. 12.2 Genre, parallel species n° 1, parallel species n° 2, parallel species n° 3

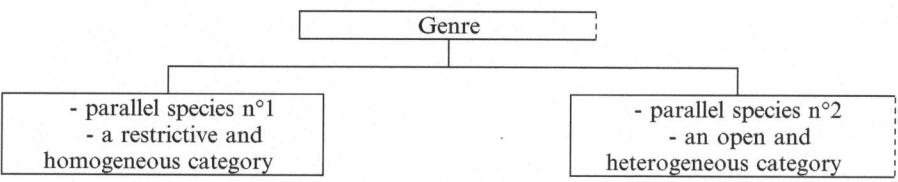

Fig. 12.3 Genre, parallel species n° 1, parallel species n° 2

is a homogeneous category that fulfils the function of completeness, whereas the unspecified contract is a heterogeneous, undefined notion that fulfils the gateway function (see Fig. 12.3).

This method will thus enable us to propose new contours for the legal identity and the basis in statute of the service contract.

12.3.2 A New Classification for the Service Contract

A brief presentation of a complete classification of contracts. In carrying out our classification, a "division-separation," we start with the contract, the most generic category of the branch, and we divide it into triads that divide in turn into other triads. More than the contract itself, we are in fact dealing with the different purposes of the contract, that is, the economic operation aimed at and wished for by the contracting parties. Indeed, a contract, whatever it may be, is no more than an instrument, an implement serving an economic operation. In this regard, what characterizes and distinguishes it from another contract is the economic operation that it facilitates.

The divisions of this contractual purpose, i.e. the economic operations aimed at, may concern either structure (a trading or organizational structure) or the elements brought together by this structure (services, situations . . .).

Starting with the contract, the most generic category of the branch, and dividing it seven times, we are able to arrive at the service contract.

The service contract must not be undefined. As we have just stated, in order to arrive at the service contract, we divide the most generic category of the classification into a series of subdivisions that allow us to discover new species and subspecies of the contract, at each hierarchical rank.

First, regarding the contract, we can state that it is a catch-all category catering for all possible individual aims, at least, all those not proscribed by law. The contract is a tool, a process for bringing about individual objectives. At the hierarchical rank of the contract, the purpose aimed at is totally undefined, indeterminate. The contract is therefore an essentially indeterminate and non-predicative notion, devoid of any attribute or positive definition. The indeterminate character of the most generic category of the contract is necessary since it guarantees that all future iterations of the practice will be integrated into the classification. The contract is a living

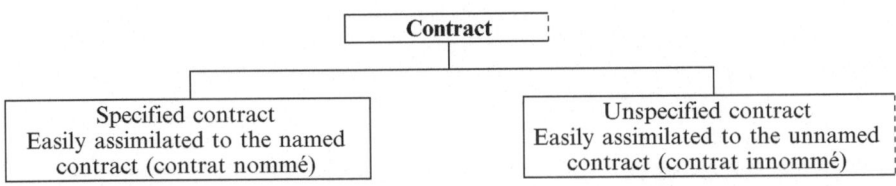

Fig. 12.4 Contract, specified contract, unspecified contract

instrument, evolving with the needs of economic actors. The indeterminate nature of its genre does not, however, prevent us from clearly identifying and determining some of its subcategories.

In order to reconcile the need to classify defined categories on the one hand, and on the other hand, the need to integrate categories that are as yet undefined, we must subdivide the generic contract into two subcategories: the defined and the undefined contract. This first subdivision is obviously close to the one in positive law and provided for in Article 1107 of the Civil Code, between the named and the unnamed contract. If indeed there is a difference between our prospective subdivision and that of positive law, it is mostly a terminological one. Certainly we think that in the perspective of an Aristotelian classification, the terminology "specified contract" is better than "named contract" since what counts most, from this perspective, is the definition of the contractual concept, not its name (see Fig. 12.4).

Since we wish to give to the service contract a defined, determinate, identified nature, in our prospective proposal it must necessarily be included in the category of defined contracts and no longer compete with the category of undefined, non-predicative, i.e. unnamed contracts, as it does in positive law. This first option marks a first break with the solutions proposed by positive law.

The service contract must be included in the category of the exchange contract, excluding that of the matrix contract. Since our reasoning is underlain by the desire to discover a positive nature in the service contract, the second subdivision of the contract we shall focus our interest on, is the subdivision of the specified contract.

We have opted for a first subdivision of the specified contract, as defined according to the specificities of the structure of its contractual purpose.

This subdivision, contrary to that proposed in the higher rank, will appear as a triad, i.e. three parallel species that, together, cover the entire range of their common genre. This triad, like all others, consists of two extreme opposites and an intermediary position. The two extreme opposites of this triad, the exchange contract (*contrat-échange*) and the organization contract (*contrat-organisation*) are concepts that were borrowed from Didier [4, p. 635, 5]. Their hybrid, the matrix contract (*contrat-souche*) takes its terminology from Carbonnier [3, n. 138] (see Fig. 12.5).

In terms of contractual purpose, the exchange contract has a simple structure, an exchange between two symmetrical elements that are still completely indeterminate at this hierarchical rank. The only specification concerning these elements is that

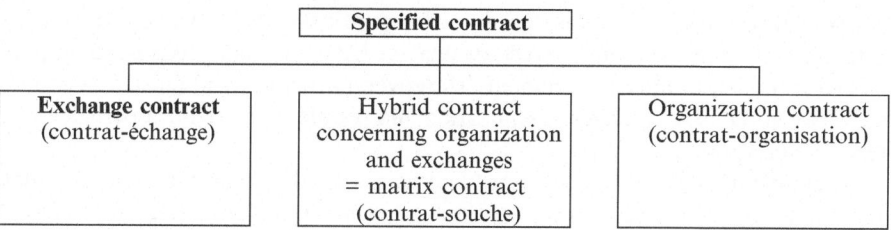

Fig. 12.5 Specified contract, exchange contract, matrix contract, organization contract

they must, of course, participate in the legal trade. The exchange category may be illustrated by the service contract or contract of sale.

Opposite the exchange contract, we find the organization contract. Concerning its purpose, this contract has a unifying structure, an interwoven, aggregated structure; in short, a very complex structure, the aim of which is precisely the bringing into being of an institution or organization. The only immediate purpose of this contract is the creation of such an organization or institution. Nonetheless, we must not delude ourselves. Organization contracts or institution contracts are only created with a view to then enabling the organization to carry out exchanges. Nonetheless, the purpose of exchange is a secondary purpose that has no direct part in this contract's contractual purpose. This category may be illustrated by the company contract.

Between the exchange contract and the organization contract, we can find the matrix contract. Its purpose is a hybrid of the goals of the two preceding ones. The characteristic of its purpose is to serve as a matrix, a framework that can be adapted and upon which lays a whole series of exchanges.

We will look more closely at two facets of this hybrid contract: first, its organizational aspect. Indeed, the contractors aim at the durability of their relationship – duration in time is then a consequence of this first facet – with a view to carrying out a series of exchanges, the contents of which are still uncertain on the day when the matrix contract is concluded. The coming into being of this inchoate organization makes it clear that the two contractors are pursuing a common interest. However, the lasting relationship established between the two contractors cannot be understood as the creation of an organization whose structure is as sophisticated as that of a business or a company. Besides, this relationship aimed at exchanging is not considered a legal entity.

This hybrid has – as regards its contractual purpose – a second facet, closer to the purpose of the exchange contract: carrying out a series of exchanges between the co- contractors, the contents of which are partially or totally undefined when the contract is concluded. The matrix contract is, in fact, an initially incomplete contract.

The distinction between the matrix and organization contracts comes from the fact that in the matrix contract, the contractual purpose includes carrying out the exchanges. The series of exchanges is carried out between the signatories to the matrix contract for as long as the agreed organization lasts between them. By contrast, in the organization contract, exchanges have a mediate purpose, a purpose distinct from that of the contract itself. Moreover, exchanges will mostly be carried out between the newly created organization and people who are third parties to the organization contract.

What distinguishes the exchange contract from the matrix contract is that, contrary to the exchange contract, the matrix contract is incomplete on the day it is concluded. At this precise moment, the parties cannot yet determine the contents of the goods and services that will be exchanged while the contract lasts. In other words, the goods and services that will be exchanged over the duration of the matrix contract are not determined on the day the contract is concluded, nor can they be determined then.

The indeterminate character of the future goods and services provided and the initial incompleteness of the matrix contract presuppose that the contractors will provide for and organize the way in which decisions will be made as to the determination of the goods and services to be carried out as long as the contract lasts.

In order to understand fully what we mean by matrix contract, let us give an illustration. Contracts known as 'integration contracts' belong to this category. An integration contract is a contract in which a contractor joins the organization of the other contractor with a view towards exchanging services with him. A typical example of this type of contract is the employment contract. In this category, the power of deciding what material services the worker will provide his employer with belongs to the employer only. Equally in this category, we find the contracts known as framework contracts. In a framework contract, one contractor supports the organization of another in order to exchange services. The organizations are not completely integrated. An illustration of this contract can be found in the set of distribution contracts enabling a franchisee or concessionary to be supported by the organization of the person who grants the franchise or concession. In this type of contract, since the four decisions made by the plenary assembly of the Court of Cassation on December 1st 1995 (Bull. Civ., n. 9, Revet, 1977, pp. 37–x) a power of unilateral decision has been recognized to the benefit of the contractor who provides the goods and services for the duration of the contract. These decisions effectively confer the right to set the price unilaterally.

As we have just stated, the matrix contract category must be clearly distinguished from that of the exchange contract, but also from that of the service contract. Indeed, we think that the service contract has a simple structure clearly set out in Article 1710 of the Civil Code. In that respect, it must be considered a true exchange contract and it must follow the same rules. It must not, on the contrary, be included in the hybrid category of the matrix contract or follow its rules. As we have just stated, the matrix contract is much more complex than the exchange contract and the business contract. As a consequence, there is no need in the exchange contract or in the service contract to regulate the decision-making process of either of the co-contractors while the contract lasts, whereas this need exists for the matrix contract.

So, as soon as an organizational aspect appears in a contract concerning the supplying of goods and services, the category of exchange contract and more particularly that of service contract no longer apply. An organizational aspect means we are dealing with a matrix contract. This detail represents a second difference with the contents of positive law. Indeed, in positive law, many examples of the matrix contract are inserted by default into the service contract category because of its general character which, being largely indeterminate, brings it close to the unnamed (unspecified) contract.

After studying the genre of the exchange contract and its two parallel genres, we will now mention a few subdivisions between the exchange contract and the service contract, in order to study the genre that directly includes the service contract and its two parallel genres.

The service contract must not be a contract of supply whether definitive and temporary or continuous and temporary.

The genre that directly includes the service contract, and to which we gave the name of "isolated/unique and definitive contract of supply" (*contrat de fourniture ponctuelle et definitive*), has two parallel genres: the continuous and temporary contract of supply (*contrat de fourniture continue et temporaire*), its extreme opposite and the temporary and definitive contract of supply (*contrat de fourniture temporaire et definitive*), and their hybrid (see Fig. 12.6).

The *isolated/unique and definitive contract of supply* is about handing over, definitively, once, and once only, something to be used. The service contract, as we have just stated, is a typical example. The same applies in the domain of sales.

The *continuous and temporary contract of supply* consists in providing goods and services continuously and temporarily, i.e. the establishment of a situation so useful as to be paid for. In this second extreme opposite, the establishment of a temporary situation of use along with protection and guarantee is at the heart of the contractual purpose. A lease (paying a rent to use something) and the insurance contract are examples of this.

Between these two contracts, there is a hybrid one: *the temporary and definitive contract of supply*. This contract has a dual purpose: on the one hand, the definitive handing over of goods or services; and on the other hand, the establishment of

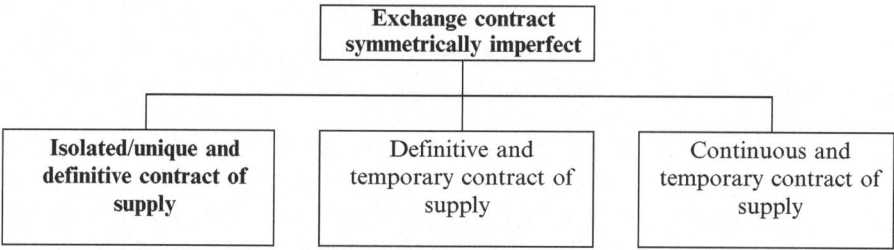

Fig. 12.6 Exchange contract, isolated and definitive contract of supply, definitive and temporary contract of supply, continuous and temporary contract of supply

a temporary situation, since the provision of such goods and services implies duration. This hybrid might seem paradoxical because of the clear opposition between temporary and definitive provision of services or goods. Yet, in practice, this paradox has been overcome.

A first example of this contract is the subscription contract which can be analysed as the definitive and gradual provision of multiple goods and services. In this contract, over a given duration, a series of successive transactions is carried out. In this respect, we should add that this contract for definitive and gradual supply is not to be mistaken for the matrix contract. The temporary and definitive contract of supply is essentially a complete contract and is destined to remain so, in contrast to the matrix contract. In the gradual and definitive contract of supply, the goods and services to be provided are determined at the outset then yielded little by little. In contrast, the matrix contract is, at its inception, essentially incomplete as well as partially indeterminate. The need to determine the goods and services to be provided over the duration of the contract implies unilateral or bilateral decision making (exchange of consent) as the contract is being carried out, in addition to the initial agreement to determine the contents of the goods and services.

We can give several other illustrations of the definitive and temporary contract of supply. For instance, the lease contract is a temporary contract of supply that may become definitive. In this contract, a temporary situation of use may be followed, when it ends, by a definitive concession; A loan for consumption (e.g. the loan of money) means both a temporary and definitive supply because of the fungibility and the consumable character of the property loaned. Indeed, for the lender to put the loaned object in the hands of the borrower implies the definitive transfer of this object into the latter's hands. Yet this situation is only temporary, because the borrower has in the end to return the object or at least its equivalent.

It is important to add that the service contract is not to be counted as either a continuous and temporary contract of supply or as a definitive and temporary one. Indeed, the service contract, because of its simple structure, can be analysed only as an isolated/unique and definitive contract of supply.

The service contract is a direct species of the isolated/unique and definitive supply of goods and service contract. As we have just stated, the service contract is an isolated/unique and definitive contract of supply. It is even the direct species of this genre, as in the case of sales, whether the sale implies the provision of goods that are immediate or deferred with regard to the date of conclusion of the contract.

The purpose of the genre of isolated/unique and definitive contract of supply can be analysed as the definitive surrendering to the client of a one-off provision of goods or services for a price. The notion of definitive surrendering implies the transfer, relinquishing, or concession of the goods and services to the client. Although it is close to the notions of property transfer/conveyance and appropriation, the notion of definitive surrendering must be distinguished from them. There is no complete identity between these concepts. Not all the goods and services thus surrendered are capable of being appropriated. This observation comes with two corollaries.

On the one hand, only those isolated and definitive contracts of supply that concern goods and services that can be appropriated (such as a building), imply the transfer of the ownership of that thing to the client. Appropriation of the good through a property transfer is only a secondary effect of the isolated/unique and definitive supply contract and also merely one of the possible effects.

On the other hand, transferring property must no longer constitute, as in positive law, a criterion for distinguishing between service contract and contract of sale nor a criterion to designate the contract of sale and exclude the service contract. In our prospective proposal, the contract of sale, whether it implies deferred or immediate supply with regard to the date of conclusion of the contract, must become a contract possibly implying the transfer of property, just like the other isolated/unique and definitive contracts of supply. This detail implies that sale, as we understand it, no longer necessarily concerns things/goods that can be appropriated, but may also concern services that cannot be appropriated. In the same way, the service contract, in accordance with its genre, must also become a contract possibly implying the transfer of property. In opposition to positive law, which never considers the service contract as implying a property transfer, the service contract in our prospective proposal must imply property transfer whenever it concerns specific services that can be appropriated. It must remain as it is, with no possibility of transferring property when it does not concern services that can be appropriated in the eyes of the legal world.

The service contract must include all contracts dealing with the isolated/unique supply of specific work for a price. By subdividing the isolated/unique and definitive contract of supply we arrive at the new service contract and its two parallel species: the contract of immediate sale and the contract of deferred sale (see Fig. 12.7).

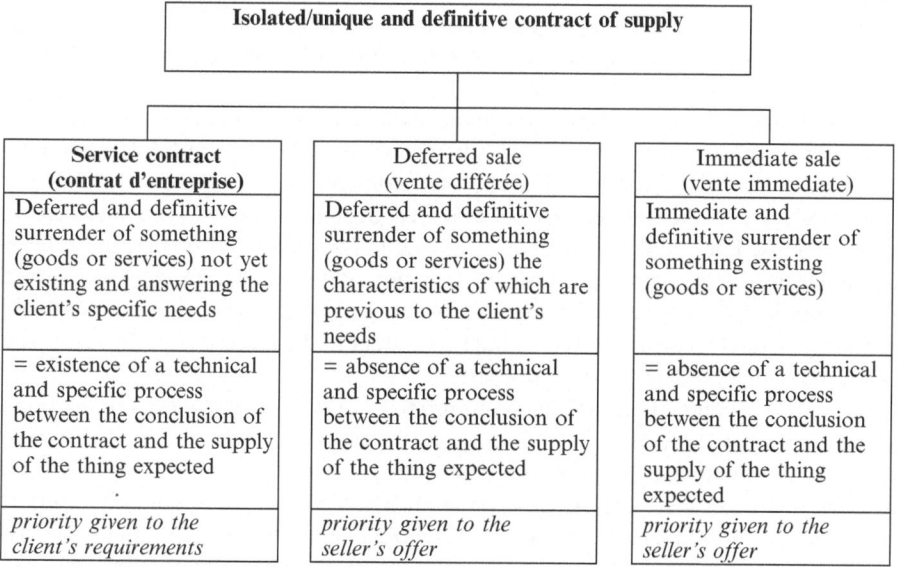

Isolated/unique and definitive contract of supply		
Service contract (contrat d'entreprise)	Deferred sale (vente différée)	Immediate sale (vente immediate)
Deferred and definitive surrender of something (goods or services) not yet existing and answering the client's specific needs	Deferred and definitive surrender of something (goods or services) the characteristics of which are previous to the client's needs	Immediate and definitive surrender of something existing (goods or services)
= existence of a technical and specific process between the conclusion of the contract and the supply of the thing expected	= absence of a technical and specific process between the conclusion of the contract and the supply of the thing expected	= absence of a technical and specific process between the conclusion of the contract and the supply of the thing expected
priority given to the client's requirements	*priority given to the seller's offer*	*priority given to the seller's offer*

Fig. 12.7 Isolated and definitive contract of supply, service contract, deferred sale, immediate sale

We have subdivided the definitive and isolated contract of supply according to the way the goods and services are provided. Indeed, what counts in the isolated and definitive supply of goods and services is, of course, the existence of the goods and services and their actual transfer to the client. Yet, the way the goods and services are provided is a still more important element because, behind this notion, fundamental notions such as *usefulness, interest and actual use of the expected goods and services* are implied. These notions are at present, in our opinion, much more important than notions of property appropriation and transfer that we consider simply as an accidental side effect of the group of definitive and isolated contracts of supply.

In our opinion, the way the goods and services are provided depends on two parameters:

– The moment when the goods and services are actually provided: immediately on the conclusion of the contract or at a time subsequent to this conclusion;
– The identity of the contractor who first stated the characteristics of the goods and services provided: the client (priority to the client's requirements) or the supplier (priority to the supplier's offer).

By combining these two parameters, we may discover three species of the genre of isolated and definitive contract of supply: the service contract and contract of immediate sale that are extreme opposites, and the contract of deferred sale which is the intermediary position.

Immediate sale, the first extreme opposite, concerns the immediate provision of goods or services, a provision of something that necessarily exists and the characteristics of which are contained in the supplier's offer.

The service contract, the second extreme opposite, concerns the necessarily deferred provision of goods and services that are to be determined according to the client's special requirements/needs.

The contract of deferred sale, the intermediary position, concerns the deferred provision of goods and services, the characteristics of which are determined by the seller. What differentiates this contract from the service contract is that the characteristics of the services to be provided depend in no way on the client's requirements. For example, the sale of a standardized material object, the sale of a standardized trip, or of standardized transport is to be described as a deferred sale. This is also the case of a mandate when it is understood as a standardized service provided by an agent to conclude a contract in the name and on behalf of the principal with a predetermined third party.

The service contract, according to our proposal is, unlike the contract of the same name in positive law, of a more homogeneous nature. It is the contract in which the contractor commits himself to implementing a specific technical process for a price in order to satisfy the client's requirements. The contractor commits himself definitively to surrendering services of a specified nature to the client; that is to say, to answering specifically the client's requirements at a date necessarily later than that of the conclusion of the contract.

The positive and particular characteristic that restores to this contract the homogeneity it does not have in positive law is the specific character of the technical process that has to be implemented in order to satisfy the client's requirements. In fact, the new criterion that we propose for the service contract is a generalization of the specific work criterion applied in domestic French law, (excluding consumer law) only to those service contracts which concern a tangible, movable item to be manufactured or to be produced.

This generalization of the specific work criterion calls both for re-inclusions and exclusions. The new service contract first includes the provision of any specific services or products for a price.

In this first respect, the scope of this new contract is indeed narrower than that of the service contract (*contrat d'entreprise*) in positive law. Indeed, in positive law, all services, including standardized services the characteristics of which are predetermined by the provider, can only be provided through a service contract (*contrat d'entreprise*). In our proposal, on the contrary, providing standardized services, just like providing future standardized items, must belong to the category of deferred sale. This moving of the present boundary between service contract and contract of sales is only possible, however, because we have accepted the relegation of appropriation and its corollary: the transfer of property, to a secondary, merely accidental role in the group of isolated and definitive supply contracts.

This solution is but an extension of the ideas of René Savatier who, as early as 1970, showed that sale in its economic sense could also concern services that cannot be appropriated and not only material objects that can.[2]

This idea, moreover, is not new in positive law. Indeed, European law used the word sale in a case of package trip provision, in a recent European directive dated 13th June, 1990, transposed into French law in a law dated 13th July, 1992 itself completed by a decree dated 15th June 1994.

Like the above mentioned sources, we think that the provision of standardized services must be included in the category of deferred sale and not that of the service contract.

The generalization of the specific work criterion also calls for re-inclusions. Indeed, in positive law, the specific work criterion is excluded as a contract for the provision of movable material services. This, in French consumer law, comes from European law and in international law. In our proposal, providing tangible, movable items to be manufactured or produced, if this is considered as specific work, must in all cases be deemed a service contract and not a contract of sale. In virtue of this aspect, the scope of the new service contract is wider than that given by positive law.

The three subspecies of the service contract can be distinguished from one another according to how difficult it is to satisfy the client's requirements. The service contract, as we propose it, can itself be divided into three contracts: the

---

[2]Cf. Savatier [6]: "Dans le langage de [l'Economie politique, la statistique et la comptabilité], qui sont, par rapport au droit, complémentaires, le mot 'vente' est, en effet, utilisé pour tout échange, fait contre de l'argent, d'un bien économique quelconque, par un échangiste indépendant".

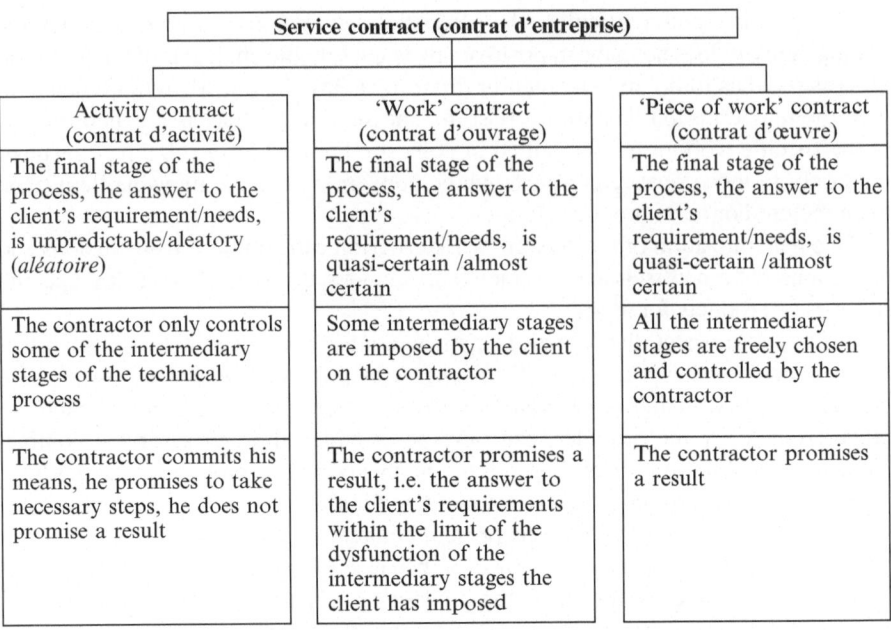

Fig. 12.8 Service contract, activity contract, work contract, piece of work contract

activity contract (*contrat d'activité*), the 'piece of work' contract (*contrat d'oeuvre*), in which the contractor controls all the elements of the work to be carried out, and the 'work' contract (*contrat d'ouvrage*), in which some elements are imposed on the contractor. The criterion for distinguishing between these three contracts is based on how likely it is that the goal aimed at by the specific technical process is achieved (see Fig. 12.8).

The activity contract is the service contract that offers the lowest degree of likelihood of obtaining the final goal, that is: the answer to the client's requirements, since the contractor can only promise intermediary stages of the process: his means, his activity, but not a result.

The medical care contract or the teaching contract are contracts in which an agent has to find a contracting party for his client. Whether called a commission or brokerage contract, these can be considered illustrations of this contract.

The 'piece of work' contract (*contrat d'oeuvre*) is the service contract that offers the highest degree of likelihood that the final stage of the process will be obtained, that the client will obtain the answer he required. This is because the contractor provides and controls all the elements of the technical process. He therefore promises that final stage of the technical process, the result of his activity, i.e. the implementation of his means. An order for an object to be created, such as making a piece of furniture, where the client does not impose on the contractor any material or any design activity, or a contract to build a specific building on land belonging to the provider can be deemed 'piece of work' contracts.

The 'work' contract is the intermediary position in this triad. The contractor promises a result for his activity just as in the 'piece of work' contract. Yet this promise is subject to the limits of dysfunction of the intermediary stages in the process that have been brought and imposed by the client (material, conceiving an answer), as the contractor does not know or control them perfectly. A contract for a piece of furniture to be made out of material imposed by the client on the contractor or a contract for the specific construction of a building on land owned by the client are examples of 'work' contracts.

The new classification and boundaries of the service contract are valuable in so far as they allow us to identify an authentic legal framework for that contract and its three subspecies.

12.3.3   The New Legal Framework of the Service Contract and Its Three Subspecies

A brief outline of the legal framework of the service contract. The fundamental features of the *legal framework* of the service contract are a consequence of the two fundamental features of its nature:

- The deferred character of the provision of services
- The specific character of the services tailored to the requirements/needs of the client

The legal foundation for the service contract is based on the priority given to the request and the specific requirements of the client, as well as on the correlated necessity for the contractor (entrepreneur) to implement a process with a specific technical purpose before he can provide the services to the client. This necessarily calls for two successive stages of encounter for the expectations of the contracting parties.

(a) On the day the contract is concluded, the first stage of encounter, the services do not yet exist and can most of the time only be determined in reference to the client's needs. The payment of the contractor/supplier is not then necessarily to be determined on that day, simply because the client cannot foresee or control any shortcomings in the promised services prior to the contract as he would be able to in the case of an immediate sale.

(b) Since the service must be created before it can be provided through the implementation of a process with a specific technical purpose, the contractor is responsible for many positive obligations:

  - A dual principal obligation: to carry out specific work (i.e., to implement a process with a specific technical purpose) consistent with the client's needs and respecting the prevailing rules

- A possible and polymorphous secondary obligation: to provide the service as soon as it is ready or to return the material belonging to the client which was used to deliver the service

(c) Non-obligational effects also exist:

- A possible property transfer/conveyance (where the services can be appropriated) that is carried out either on the day of the provision of services, or on the day of the integration of the services into the item or property of the client
- A systematic risk transfer after the promised services have actually been delivered

(d) With the provision of the service, there appears a second stage where the two sets of expectations should coincide. The client must immediately check the services and look for possible shortcomings. At this point, he may choose between three possibilities:

- To refuse to accept
- To accept without any reservations
- To accept with reservations

If the two parties did not come to an agreement about the price when they first stated their expectations, an agreement on the price must also be found after the contractor has effectively provided the service.

(e) To accept with or without reservations renews the dual and principal obligation of the contractor (suitability of the contractor's answer to the client's requirements, plus respect of the regulations in force at the time) and makes it a single test obligation (*obligation d'épreuve*), i.e., a guarantee that the services respect standard prevailing rules throughout the reasonable period of their provision.

This implies that the suitability obligation of the contractor expires on acceptance, with the exception of any failings about which reservations had possibly been expressed.

The new test obligation (*obligation d'épreuve*) is distinct from a guarantee from hidden defects, linked to the idea of wilful misrepresentation, as well as from the notion of false declaration concerning the functioning of the services provided. It originates rather in the obligation on the contractor for skill, efficacy, or ability. The test obligation concerns breaches of standard prevailing rules, not only concealed, but also apparent and about which no reservations were expressed on acceptance. This advantage favouring the client can be explained by the fact that the contractor, whether he is a professional or not, demands a price for his activity. He then necessarily has an obligation of competence which he must not be able to get around by means of any contractual terms.

A brief outline of the legal framework governing the three subspecies of the service contract. This framework is simply a refinement of the rules that apply to the service contract and formalizes how difficult it may prove to obtain the promised services.

An activity contract, for instance, a medical care contract, is a contract in which the contractor promises only to employ the necessary means and to take the necessary steps, but does not promise the result that represents the answer to the client's specific need. The specificity of this contract implies it should be regulated according to three and not two procedural stages: three times when the two minds of the contracting parties should meet.

The first occasion occurs as soon as the contractor undertakes to implement all the means at his disposal to meet the client's specific requirements. This first agreement concerns a provision of material services that can be defined only according to the client's specific requirements and at a yet undetermined price. This first stage creates an obligation for the contractor to provide a diagnosis.

Once the diagnosis is given, there comes a second procedural stage. This second stage is in addition to the stages that exist for the service contract.

This procedural stage begins either with the duty of modesty on the part of the contractor, he will then retract his initial consent because he feels he cannot provide the promised services, or with the duty for him to warn the client against the risks of the contract at his expense. This duty to inform is equivalent to the confirmation, the ratification of his initial consent. After this warning, the client may either refuse or repeat his initial consent.

If he repeats his initial consent, this creates a new principal obligation for the contractor to provide specific work that can be understood as a reinforced obligation of means.

Once the services have been provided and delivered into the client's hands, there comes a third procedural stage, the tacit reception of the provided services and the express acceptance by the client of the price proposed by the contractor.

The extreme opposite of the activity contract, 'piece of work' contract, such as a contract concerning an order for a work of art, is a contract by which the contractor promises a result, i.e., the answer to the client's requirements. Unlike the activity contract but like the service contract, this contract implies a dual exchange of consents.

This contract's principal specificity fosters a principal obligation to provide adequate services, which can be understood as an obligation of results.

The intermediary position between the activity contract and 'piece of work' contract, the 'work' contract, such as a contract for the repair of movable goods, is a contract by which the contractor promises a result, the answer to the client's requirements, within the limits of possible dysfunction of the elements imposed by the client (materials, vision of the answer). This contract, like the service contract, requires a dual exchange of consents.

The two principal specificities are the following. On the one hand, the obligation of modesty and that of information that lie with the contractor essentially concern the intermediary stages imposed by the client (materials, conception of an answer to the requirements). On the other hand, the principal obligation on the contractor is one of results, within the limits of possible dysfunction of the predetermined intermediary stages.

The proposal made here of a new legal identity, indeed a new statutory framework for the service contract, could be the means to remedy the lack of a defined nature that currently characterizes the contract for the provision of services for a price. Our proposal could also put an end to the incoherence that results from a contract that is fundamental at an economic level is subject to almost no legal regulations.

References

1. A. Bénabent, *Les contrats spéciaux civils et commerciaux*, 7ème edn. (Domat Montchréstien, 2006)
2. R. Blanché, *Structures intellectuelles*. Librairie Philosophique (J. VRIN, Paris, 1966)
3. J. Carbonnier, *Les Obligations, Droit civil* (PUF, Thémis, 2001)
4. P. Didier, Le consentement sans l'échange: contrat de société. Revue de Jurisprudence Commerciale. **11**, 74–80 (1995)
5. P. Didier, Brèves notes sur le contrat-organisation, in *L'avenir du droit, Mélanges en hommage à Fr. TERRE*. (Dalloz, PUF, Edition du Jurisclasseur, Paris, 1999)
6. R. Savatier, La vente de services. Dalloz 1971, Chronique 223–231 (1971)
7. J. Sénéchal, Le contrat d'entreprise – problèmes de qualification, approche française, in *Droits des contrats France*, ed. by B. Tilleman, A. Verbeke, P.-Y. Verkindt. Belgique, vol. 2 (Larcier, Bruxelles, 2006), pp. 25–31
8. J. Sénéchal, Le contrat d'entreprise au sein de la classification des contrats spéciaux : recherche sur un double enjeu du mouvement de recodification du droit des contrats. PhD Thesis, Presses Universitaires d'Aix-Marseille, Aix-en-Provence, 2008

ERRATUM

Chapter 5
Suspensive Condition and Dynamic Epistemic Logic: A Leibnizian Survey

Sébastien Magnier

© Springer International Publishing Switzerland 2015
M. Armgardt et al. (eds.), *Past and Present Interactions in Legal Reasoning and Logic*, Logic, Argumentation & Reasoning 7, DOI 10.1007/978-3-319-16021-4

DOI 10.1007/978-3-319-16021-4_13

Reference [17] has been mistakenly placed at the end of reference list instead of placing at number [7]. This reference is now placed correctly in the reference list.

The online version of the original chapter can be found at
http://dx.doi.org/10.1007/978-3-319-16021-4_5